IRISHOGRAPHY

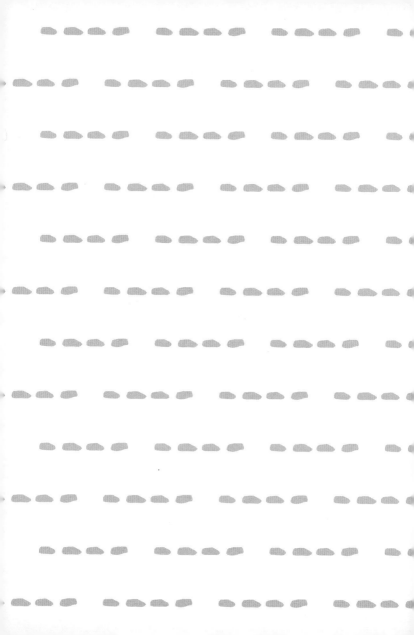

IRISHOGRAPHY

CONNEMARA, CROAGH PATRICK, COPPERS AND EVERYWHERE ELSE WE LOVE IN IRELAND

RONAN MOORE

GILL BOOKS

GILL BOOKS
HUME AVENUE
PARK WEST
DUBLIN 12

Gill Books is an imprint of M.H. Gill & Co.

© Ronan Moore 2016

978 0 7171 7121 7

Edited by Ruth Mahony
Designed by Fidelma Slattery
Illustrations by Fuchsia MacAree
Printed in Grafo, Spain

This book is typeset in Daft Brush and Gotham.

The paper used in this book comes from the wood pulp of managed forests. For every tree felled, at least one tree is planted, thereby renewing natural resources.

A CIP catalogue record for this book is available from the British Library.

5 4 3 2 1

For our baby girl, Esme,

who brings us so much joy, x

ACKNOWLEDGEMENTS

There are many people I would like to thank for their help, tips, advice and insights. These include, in no particular order: Frances, David H., Anne-Marie and Richard, Paraic, Deirdre, Elle, Grainne, Carol, Seamus, Brendan, Mairead, Frances F., Mark, Mags, Mary, Gordon, Charlotte, Melissa, Peter, Mick, Pauline, Daniel, Darran, Maria, Fintan, Ian, Podge, Camilla, Leo, Noel, Rachel, Tracy, and, of course, Mam and Dad.

Finally, I'd like to acknowledge all the hard work of those at Gill Books for the endless hours of support and assistance.

CONTENTS

LEINSTER

CARLOW

Carlow is a little like the Ghana of the south-east.

This is not due to their shared history of slavery, sugar beet, golden beaches, flowing rivers, beautifully interwoven Kente cloth, agricultural shows and rich pre-colonial Ashanti rule. No, instead it is due to the fact that Carlow and its West African cousin both have the same animated hues of red, green and yellow on their football jerseys. These shades make the Carlow shirt the most vibrant in the land – and almost as colourful as their annual Championship first-round exit.

Despite this annual sporting setback, and despite the fact that they are the second smallest county in Ireland and have the very inauspicious nicknames of the Scallion Eaters and the Fighting Cocks, Carlow seems a county very much at ease with itself. Perhaps this is due in part to its beautiful balance of rural and urban eastern life and in part to the knowledge that for thirteen glorious years in the 14th century, Carlow town served as Ireland's capital.

BROWNSHiLL DOLMEN

dolmen

/ˈdɒlmɛn/

Noun

1. A megalithic tomb with a large horizontal stone laid on upright ones.

IDiOT'S GUiDE TO BUiLDiNG A DOLMEN

Step 1: Gather a number of boulders (not to be mixed up with rocks) and stand them close to each other to make what will be the legs.

Step 2: Gather a really big boulder and, with the help of a few mates, rest it on top of the legs to form a capstone.

IDiOT'S GUiDE TO BUiLDiNG BROWNSHiLL DOLMEN

Step 1: Gather a number of boulders (not to be mixed up with rocks) and stand them close to each other to make what will be the legs.

Step 2: Gather the biggest feckin' boulder you've ever seen and with every man, woman and child in the community, place it on top of the legs to form the mother of all cap-stones!

While Brownshill Dolmen (or the Kernanstown Cromlech, as it's sometimes known) may not be as immediately recog-nisable or as stately as the likes of Poulnabrone Dolmen in County Clare, it is every bit as fascinating. Beautiful in that 'only his mother could love him' sort of way, Brownshill Dolmen is the colossus of Irish portal tombs – or any portal tombs for that matter. What makes Brownshill unique is its capstone, which, at over 103 tonnes, is the heaviest cap-stone in Europe.

Visiting Brownshill Dolmen often leaves a tourist with two questions. The first: where on earth do you find a 103-tonne boulder? It's not usually something you dig out of your back garden with a stick. The second: how on earth did a group of farmers about 5,000 or 6,000 years ago find the time and ability to lift a 103-tonne capstone – a weight roughly similar to that of a Boeing 757?! Whatever the answers, I just really hope they kept their backs straight and bent their knees while they were doing it.

THE TEMPLE OF ISIS

It must come as a shock to recently married young couples moving to the picturesque village of Clonegal, nestled in the Blackstairs Mountains of east Carlow, to find out that alongside the local schools, pubs and church is their very own Temple of Isis!

Thankfully, rather than the bad ISIS hell-bent on world domination, Clonegal's is a good Isis, inspired by the Egyptian goddess of the same name. Free to join and open to all, the Isis fellowship of Clonegal was founded as a peaceful world religion that promotes the female aspect of divinity.

Finally, while the bad ISIS's annual plans generally begin with spreading murder and doom, the annual plans of Huntingdon Castle's residents, where the Temple of Isis has been located since its establishment in 1976, are a little bit more positive. They usually begin with snowdrops, lambing and woodland walks, with absolutely no blowing up of ancient monuments or beheadings permitted.

THE RIVER BARROW

The River Barrow is sometimes more famous for being a member of the Three Sisters, a multi-platinum-selling all-girl river band styled on the Corrs except without a brother who doesn't believe in Santa. However, it has achieved considerable success in its own right.

Though the Barrow, the country's second longest river, rises in Laois and flows out from Waterford, it is perhaps most synonymous with Carlow, through which it flows and along, from its northern to its southern tip. And it is down its Carlow section you'll find arguably the most enjoyable barging routes in the country.

Of course, this barging is not the I-got-out-of-bed-late, now-I-am-going-to-miss-my-plane, my-life-is-that-little-bit-more-important-than-yours-so-you-don't-mind-if-I-jump-the-queue type of barging but the other form, that slow steering of canals and rivers. The Barrow Navigation begins in Athy, Kildare, where the Grand Canal meets the River Barrow; from there, it ambles down through County Kildare. Once it reaches the county of Carlow, it then takes on its characteristic meander down the river through the stops and sights at Milford, Leighlinbridge, Bagnelstown, Goresbridge, a flirt in Kilkenny at Graignueamanagh before finishing at St Mullins, all the time flanked on both sides by rolling countryside and over-hanging trees.

In what is sometimes called one of the most beautiful navigable stretches in Europe, time slows down, the world gets quieter and as long as you can remember to tie up your barge when you drop into a local pub for a quick one, it really is a nice way to spend a week.

ALTAMONT GARDENS

Q. If it takes one man with access to a horse and cart a day to dig and remove one square metre of earth, how long will it take 100 men to do the same for 10,000 square metres?

A. Thanks to the centrepiece lake in Altamont Gardens, which is roughly this size, we know this answer to be two years.

Now, if you have ever worked in the garden, you'll know that digging a square metre of earth is hardly Herculean. However, when you consider that the lake in Altamount Gardens was undertaken between 1847 and 1848 and that the men doing this work were also enduring the Great Famine, you can probably understand why they were not operating at full tilt.

Nevertheless, they were operating, and this was in part thanks to the Dawson Borrer family who employed an army of local men, paying them the not-inconsiderable sum of £12,000 to carry out the task, money that went a long way back then.

The lake is just one part of what are Carlow's most impressive and romantic gardens. The grounds are made up of formal and informal gardens, populated by native and exotic trees, young and mature, with lawns full of roses and rich shrubbery surrounded by rhododendrons and rare trees that all flow down to the centrepiece lake. Of all the flowers, plants, shrubs and trees in Altamont Gardens, the most famous are its snowdrops: it boasts 100 varieties of these hardy but beautiful white winter warriors – which is exactly 99 more varieties than most people know exist.

DUBLIN

While Dublin is a lot more than the city – Howth Head, the Forty Foot, Terminal 2, for example – when talking of the county, it is often the metropolis that immediately comes to mind. And as the nation's capital, it is easy to figure out why.

The city was first properly established as a Viking settlement back in the 10th century. Its original name, *Dubh Linn*, ('black pool' in English), was the Gaelic term for the body of water where the Scandinavians anchored up their longboats. In 1169, the first stag party from Wales arrived, an Anglo-Norman affair led by Strongbow. Within a few decades, the city had become a cultural melting pot and the stag capital of Europe, a title it holds to this day.

A little-known fact is that at one stage in history, around 1700, Dublin was widely considered to be the second largest city in the world after London. While today it probably wouldn't make the top 500 most populated cities in the world, it is Ireland's only city with over a million residents.

Despite its size, Dublin is a remarkably friendly place and its people are renowned for being great craic to go out with. It is said that strangers to the city are more likely to receive a drink from the natives of Dublin than in any other place in Ireland. They are also more likely to be asked '*what are ye looking at?*' and '*are you calling me a liar?*' here than anywhere else too.

THE PHOENiX PARK

Take 1,752 acres of land and add:

> The residence of the Irish president

> A zoo

> A giant Christian cross (from when the Pope visited)

> The largest obelisk in Europe (a large pointy phallic-type monument)

> A small castle

> Victorian gardens

> A hospital

> Police headquarters

> The Ordnance Survey Ireland headquarters

> A plush house to host foreign dignitaries

> The US ambassador's gaff

- A fort

- 12 soccer, 7 GAA, 3 camogie and 2 cricket pitches, 1 polo grounds and a great big space for model aeroplanes

- A calendar of annual events ranging from running to motor racing and musical concerts to national homecomings

- And a team of park constables who do battle on a daily basis with their greatest nemesis: the Frisbee-playing disposable barbecue chef.

Then, cover with 30% deciduous trees and populate with half of the mammal species of Ireland, 40% of the bird species, a herd of about 500 wandering fallow deer and more picnic space than you can shake a sandwich at.

Make it free to enter, open 24 hours a day, seven days a week, 365 days a year and what you are left with is the largest enclosed recreational space within any European capital city – a place we call the Phoenix Park.

Now, if we only had the weather for it!

THE SOUTH QUAYS

A FiRST DATE GUiDE TO THE SOUTH QUAYS

Meet at the gates of Ireland's foremost university, Trinity College. While this is not on the South Quays, it is by far the best place to meet on the southside – as long as no disgruntled pensioner is trying to ram his car through it. If you fancy your chances, head for the Book of Kells inside its grounds. The queue to see it on a summer's day is the relationship equivalent to six months backpacking together through South America. If you are still talking by the time you finally reach the top of the line, you're as good as married.

(If you're coming from the northside, meet on O'Connell Street – at the Spire if your date has only moved to Ireland; under Clery's clock if they are originally from the country; or at the GPO if you're one of those people who hate it when others don't stand up during the national anthem.)

CULTURAL YOU: Head first for the cultural quarter of Temple Bar. Unless you want to give the impression that you have a small drinking problem, don't suggest joining the English stag from Lancashire for a pint. Instead, save the €100 you would have spent on a round of drinks and bring your date to any of the cultural hotspots surrounding you, from the Irish Film Institute and Meeting House Square to the Project Arts Centre and the National Photographic Archive. To help impress, be on first-name terms with someone working in a shop, preferably one that sells books, at the far end of Temple Bar who you just happen to bump into.

HISTORIC YOU: Bring them up the river to Wood Quay, where one of the most extensive Vikings ruins in northern Europe was uncovered in the 1970s and then buried in concrete when Dublin Corporation built their headquarters here. Point out the half-buried Viking ship on the pavement just up from it, which they left out as a warning to others. Mumble under your breath something about '*Charles Stewart Parnell*', '*Romantic Ireland being dead and gone*', how '*it eats you up every time*' and look emotional. Wave a fist at the building if you feel it appropriate and then lighten the atmosphere by heading up to grab some fish and chips in Leo Burdock's before visiting Ireland's oldest cathedral, Christ Church – because who doesn't want to see a mummified cat chasing a mummified mouse? Good times!

SOCIAL YOU: Head down to your final stop, the Guinness Storehouse. Resist the urge to tell your date that you brew your own craft beer. Instead, wander around the most-visited tourist destination in the country, getting to know each other and playing '*What country do you think they are from?*'. Finish it with a night out (for 'night', read 'early evening'), hanging with a group of really friendly American retirees from Utah in the Gravity Bar. Not only will your witty Irish ripostes make you seem like the friendliest person there, thus guaranteeing a second date, but you can be sure that most of those Americans are not going to drink their complimentary Guinness, which means there will be plenty of spare pints to go around. Word of warning: this only works the once before the barman gets wise to your ways so choose that first date carefully.

CROKE PARK

Every year Croke Park, or Croker as it is affectionately known, plays host to the some of the most exciting events in Ireland – and we're not talking One Direction concerts or the Eucharistic Congress. It is during the summer, when the stadium becomes thronged with colour as crowds stream in to watch their hurling and football counties do battle on this semi-sacred turf.

Despite being little known outside of Ireland, the sports ground, with a capacity of 82,300, is Europe's third largest stadium and one of sporting Ireland's great little secrets.

There are a hundred little things that people remember when they have been Croker. It could be standing on the Hill in a sea of blue or green or jealously sitting opposite in the Canal. Maybe it is high up in the Hogan or Cusack, remembering back to when a column of concrete obscured their view, or tucked in the Nally with a group of noisy school-children from out the country. For some, it is the burger and chips on Jones Road or arranging to meet up outside the Big Tree. For others still, it is the hats, scarfs, headbands and Fruit and Nut bars, *'Two for a euro!'*

Whatever the little things you might associate with Croker, it is the big thing that every year most counties hope for. That big thing begins with throw-in at 3.30pm, on the first, second, third or fourth Sunday in September, depending on whether it is football, hurling or camogie, and every supporter dreams it will be their team, their county who takes their place on the field come throw-in. And then they will hope and pray that it is their players who will walk the steps of the Hogan some 70-odd minutes later, speak a few sentences of Irish badly, hip-hip-hooray the losing side and take home the All-Ireland.

COPPER FACE JACKS

If Copper Face Jacks can't claim to be Dublin's premier nightclub, it can claim to be one of its most popular and is certainly its most famous one.

Open from 10pm until late, seven days a week, 365 days a year, with the exception of Christmas Eve, Christmas Day and, of course, Holy Thursday and Good Friday, Copper Face Jacks (or Coppers, as it's more commonly called) turns out profits that are roughly equivalent to a NASDAQ 100 company.

Coppers can be split into two eras, pre- and post-BC, with BC being ex-Dublin captain Bryan Cullen. When Bryan uttered those immortal lines *See yiz in Coppers!*' while accepting the Sam Maguire in 2011, he opened the floodgates for Dubliners to flock to this Harcourt mecca. Up until that moment, those from the capital had been largely absent. However, after BC's public endorsement, their presence soon added to what had long been a favourite of country-born city-dwellers, who would come here on the weekend, remember their own local disco back home, sing 'Maniac 2000' and get the shift.

Other interesting, though slightly fabricated, facts about Coppers:

★ The reason that Coppers is so successful is that it reinvests all its profits back into itself.

★ Due to the numbers of nurses here on a night out, Coppers is probably the safest place in Dublin after hospitals to go into labour at 3am on a Friday.

★ Only once, on 27 May 2012, has a Garda actually paid to get in.

★ The number of steps you must walk down to get into the nightclub is actually less than the number of steps you have to walk up to get out!

★ 43% of first-time male visitors to Coppers have admitted to having difficulty finding the bathroom from the main dance-floor.

★ Former Garda Cathal Jackson owns the nightclub, but no pictures exist of him on the internet. Even pictures that do exist of him are not actually him.

★ By 2035, it is thought that 14% of public-servant babies will have been conceived following a night out in Coppers.

★ During the Celtic Tiger, a man from Roscommon bought a Fat Frog here for €27.50.

★ Copper Face Jacks' Gold Member Cards are more valuable by weight than actual gold.

★ There are also Diamond Member Cards but only five of these exist at any given time. While no one actually knows who possesses these, it is believed they are currently in the hands of an ex-Dublin captain, a member of the Illuminati, yer one from the telly, a member of S Club 7 and the current US president.

★ While they might not always admit it, 86% of people who go to Coppers are looking for the shift.

★ And finally, the annual income from the Copper's cloakroom is higher than the GDP of Chad.

KILDARE

To an outsider, Kildare can sometimes come across as the county in Ireland most lacking in character and spirit. There are two reasons for this. The first is that their county jersey is absolutely colourless (a fact that gives rise to their nickname, the Lilywhites). This means it is possible to deck out a family of Kildare football supporters for less than a fiver, with a pack of white t-shirts from Dunnes. Second, at any given stage, half of the county's residents are not even technically there! Instead, they are in the process of commuting to or from Dublin.

But to those who know it, Kildare has as much character as west Kerry or north Donegal.

To best understand this, one simply needs to criss-cross the county, making sure to avoid the Bermuda Triangle of Irish traffic that is Naas. When you do this, you will inevitably end up on its back-roads and by-roads, and it is here that you will quickly discover that Kildare has more nooks and crannies than your grandmother's larder. You will find a Kildare that has bogs, hills, fens, canals and plains unlike anywhere else in Ireland; a Kildare whose history is filled with cairns and castles, follies and fairy forts, a county that is as much country as it is cosmopolitan; and a Kildare where, not 50 kilometres from Dublin, you can still find shops that double as pubs that double as off-licences that double as funeral homes that double as houses.

Kildare: like a Dublin suburb, but not.

THE IRISH NATIONAL STUD AND JAPANESE GARDENS

While these entries might appear very different to each other, the Irish National Stud and Japanese Gardens are in fact so closely linked that not only are they next-door neighbours, you cannot pay into one without paying into the other.

Both the Stud and the Gardens make ideal day-trips, though they stop just short of making it onto the 'Top Ten Places in Ireland to go on a First Date' list. This is because, although the Path of Life that runs through the stunning Japanese Gardens, complete with Stepping Stones of Exploration, the Hill of Ambition and the Well of Wisdom, is jaw-droppingly romantic, you can never be certain how someone will react to being brought to the National Stud. I mean, it is perhaps the largest state-run sex industry in the world, albeit for horses!

However, if women can get past the idea that the National Stud is a place where stallions stand round waiting to have sex and men can get over the fact that mares (or their owners, at least) are prepared to pay up to €70,000 for this privilege, it's quite fun in that horse-porn kind-of-way. Although do spare a thought for Tommy the Tease, a Connemara pony whose job it is to act like an equine version of a Barry White LP, getting mares into the mood before they are cruelly taken away from him.

THE BIG BALL OF NAAS

Despite failing in its bid to join the Taj Mahal, Machu Picchu and the Great Wall of China as one of the modern wonders of the world, the Big Ball of Naas remains one of the nation's most iconic pieces of road art.

Roughly nine metres in diameter, the Big Ball is a great big sphere that sits just off the Dublin Road roundabout in Naas and is visible to anyone who has ever travelled up or down the M7. It is covered with a myriad of road markings, which, according to Kildare County Council, 'follow and symbolise the motion of traffic on the nearby roads' and 'suggest the movement of the winds and ocean currents over the surface of the earth – a planet in perpetual motion'. Indeed 'Perpetual Motion' is its official title, which will probably seem ironic to anyone who has ever tried to join the motorway westbound off the roundabout on a Friday evening!

While this is the official explanation for the Big Ball's exist- ence, there are many local rumours about what it contains, with suggestions including a cryogenically preserved former Minister for Transport and a time capsule containing old Nokia phones.

Finally, though the Big Ball was already well known to anyone who'd passed Naas since its installation in 1996, it sprung to national prominence in 2008 when it co-starred in that Guinness advert starring Michael Fassbender. (What? A Guinness advert starring Michael Fassbender?! Drop book! Rush to YouTube!) In the advert, Michael leaves Dublin, walks westward and then swims the Atlantic to New York to apologise to his brother, either for doing the dirt with his brother's girlfriend or, more likely, wearing his brother's good shirt to the local disco without asking. But what all true Naas-onians know, however, is that when Michael walks by the Big Ball in the advert, he is in fact walking in the direction of Dublin, which we can only presume is because he forgot his passport (we've all done it).

ST PATRiCK'S COLLEGE, MAYNOOTH

'A long time ago in a galaxy far, far away', we had so many priests in Ireland that the European Union threatened to put a clerical quota on the country, like they had for our milk and butter. In fact, so rude was the health of the Catholic Church that annual inter-seminary games (basically a priest Olympics) used to take place across all the seminaries of Ireland.

Things have changed a lot since then, with a huge drop-off in numbers and the only seminary still operating the national one of St Patrick's College. Though numbers have waned, priests continue to answer the call, to spread the good word and trigger whole new generations of native people to pose that age-old question, *'who said Mass?'*

Throughout these changes, one thing has remained the same and that is the solemn but stunning surroundings of

St Patrick's College and, in particular, St Joseph's Square. Almost totally enclosed on all four sides by impressive ivy-covered walls and a front entrance that looks like it was made for a Harry Potter movie, St Joseph's Square continues to draw contemplation and invite inspiration in equal measure.

Wandering in the wonderfully well-kept gardens of St Patrick's College, now part of Maynooth University's South Campus, it is hard not to get philosophical, irrespective of whether you are a Kiltegan missionary soon to leave for West Africa or a Fresher recently arrived from Athy.

THE CURRAGH

'Aye, fight and you may die. Run and you'll live – at least a while. And dying in your beds many years from now, would you be willing to trade all the days from this day to that for one chance, just one chance, to come back here and tell our enemies that they may take our lives, but they'll never take our freedom!!!'

In some quarters, the Curragh is most famous for being the site where William Wallace uttered these words and led an army of fellow Scottish men in victory over the English at the Battle of Stirling. And then for being the site where William Wallace later led an army of fellow Scottish men in losing to the English at the Battle of Falkirk. And then finally for being the site where his friend Robert the Bruce led an army of fellow Scottish men in victory over the English at the Battle of Bannockburn. Of course, none of these battles actually took place in the Curragh or even Ireland, but were just acted out in *Braveheart* by Mel Gibson and the Irish army who just so happened to be stationed here and not busy invading any countries at the time.

That doesn't mean you can't still don a kilt, paint your face blue and white and run screaming across the Curragh's

plains. It's just it might be far more appropriate and less embarrassing for your kids to come, visit and appreciate everything else the Curragh has to offer:

- ⤳ It is a flat open plain of almost 5,000 acres and one of the oldest natural grasslands in Europe. And as long as you don't wander into the 815 acres the Irish army (who are based here) uses for practice, it is a great place to take the whole family for a walk.

- ⤳ The reason there are sheep everywhere is that the Curragh is 'common land', where sheep are free to live, love and graze wherever, whoever and whenever they like.

- ⤳ It is home to Ireland's top racecourse, the Curragh Racecourse, which hosts Ireland's most prestigious flat-race annually, the Derby. (In fact, the Irish name for the Curragh, *cuirreach*, translates as 'place of the running horse').

KILKENNY

Yes, there is the quaint medieval city, the rustic surrounds of Jenkinstown Park, the historic Jerpoint Abbey, the mysterious Knockroe Passage Tomb and the winding river wonders of the Three Sisters. But despite all that, when talking about the county of Kilkenny, there is really only one thing that you need to mention: hurling.

While there is a myriad of counties that have helped perfect the art of hurling, there is one county that is more synonymous with the sport than any other and this is Kilkenny. And there are some really good reasons for this:

••• It has won the All-Ireland hurling championship 36 times, over a quarter of the finals since records began.

••• It is the only county that doesn't enter a football team in the All-Ireland football championship. Even the counties of London and New York enter!

••• In 2009, more than one in three Kilkenny people (33,051) turned up at the hurling county club final between Ballyhale Shamrocks and James Stephens.

And there are some less 'real' good reasons for this too:

... In Kilkenny City's St Patrick's Day Parade, the LGBT float regularly objects to the presence of footballers!

... It is rumoured that if a Kilkenny child is unable to balance a sliotar by fourteen months, then they are considered personae non gratae at the county's crèches and montessoris.

... Until quite recently, known Gaelic footballers were refused communion in Kilkenny churches.

... The hurl is technically considered to be part of the body under local law and its presence is regularly sighted beside students sitting the Leaving Certificate, with parishioners as they say Confession and alongside the Bible in polling stations during general elections.

... And finally, until recently it was an offence to be caught driving a car in Kilkenny without up-to-date insurance, NCT, tax and two hurls and a sliotar in the boot of the car.

KiLKENNY CASTLE

While Kilkenny Castle was originally built in 1195 to control a fording-point of the River Nore, what made it really special was the Butlers of Ormond who moved into it two centuries later. Of all the Anglo-Norman families that came to Ireland, the Butlers of Ormond have to be the coolest.

First of all, their name was cool, which they changed to the Butlers of Ormond from Fitzwalter in the 14th century. Aside from making them sound like a sophisticated rock band, it meant they were one of the few medieval families whose name I could actually remember during history tests in school.

Second, their longevity as a family was cool. They were the Irish medieval-age equivalent to the Kennedys, the Clintons and the Kardashians all rolled into one. Politically astute, remarkably resilient and eternally popular, they somehow managed to toe that invisible line of remaining faithful both to the Crown and to Ireland and thus hung onto power and to their land from 1391 all the way into the 20th century!

Third, it wasn't just the men who were cool – the female members of the family were pretty impressive too, particularly Eleanor Butler, sister to the 17th Earl of Ormond, who in 1778 ran away with her neighbour Sarah Ponsonby. Eventually setting up home in Wales, they became known as the Ladies of Llangollen and were quite possibly Ireland's first openly lesbian couple.

Finally, while the Butlers' castle is cool, being one the most ornate chateaus in the country with its tastefully restored old rooms, long galleries, art collections, fine fireplaces, glorious ceilings and 50 acres of public parkland extending to its south-east, the Butlers' legacy is even cooler. This is because in 1967, the 24th Earl of Ormond, recognising that his family were no longer in a position to keep the castle going in its old glory and realising that there were '*already too many ruins in Ireland*', decided to give it to the Irish people so that it would continue to '*be seen in all its dignity and splendour*'. And he did all of this for the princely sum of £50!

DUNMORE CAVES

No one truly knows the origin of Kilkenny's county nick-name, the Cats. Some say it dates from the Cromwellian War, when English soldiers in the city tied two cats together to see them fight. Others say it stems from the lack of municipal agreement during the Confederate Wars, which led to boroughs of the city quarrelling with each other like 'cats', while others still suggest it is down to the sheer volume of strays cared for by the spinsters of Thomastown, Urlingford and Castlecomer.

Of all the suggestions, perhaps the most colourful is the one that derives from Kilkenny's fascinating Dunmore Caves. Here, within one of Ireland's deepest caves, lived the legendary cat Banghaisgidheach. Though a name like Banghaisgidheach would, in normal circumstances, have meant an almighty amount of bullying at school, Banghaisgidheach didn't have that problem as she was a monster cat. And it was in the Dunmore Caves that Banghaisgidheach slew Luchtigern, king of the mice, in an epic version of Tom and Jerry.

While remnants of this epic clash are no longer visible during a trip to the caves, there are a myriad of interestingly-named impressive calcite formations are for visitors to enjoy, but be warned – according to the 9th-century document *The Triads of Ireland*, Dunmore Caves are one of the three darkest places in Ireland, with reports of people bumping their shins off the corner of the bed on their way to the loo for over 1,200 years.

KILKENNY'S MEDIEVAL CITY

The medieval city of Kilkenny, aka the Marble City, is one of Ireland's top tourist destinations. What makes it somewhat unique among its peers is that those who visit Kilkenny often come for very, very different reasons. Indeed, those who wander through its medieval heart fall into five distinct categories.

1. THE STAGS AND HENS:

In 2009, Kilkenny marked its 400-year anniversary of being granted city status. It was a bumper year as it happened to also coincide with the city celebrating its centenary as one of Ireland's most popular stag and hen destinations. For a very long time, stags and hens have come to Kilkenny to either commemorate their departing singledom or celebrate their approaching marriage. In fact, so popular has Kilkenny become that, along with its stag and hen-hosting younger siblings of Carrick and Carlingford, it is believed that 76% of Ireland's bubble soccer, paintball, go-kart and clay pigeon shooting venues are concentrated within a ten-kilometre radius of these three towns, and at least one

Chinese industrial city is kept in business solely from sales of its willy-straws, blow-up dolls and penis hats that end up out and about on their streets.

2. THE COMEDY LOVERS: Back in 1994, someone had the bright idea (for 'bright', read 'bizarre') that Kilkenny was naturally placed to host Ireland's premier comedy festival because, as everyone knows, Kilkenny has a long tradition of humour that stretches back all the way to ... er ... 1994. And it worked, with the Cat Laughs now one of the most internationally-acclaimed comedy festivals.

3. THE ECONOMICS AND COMEDY LOVERS: Back in 2010, someone had the bright idea (for 'bright', read 'even more bizarre') idea that Kilkenny was naturally placed to host Ireland's premier economics and comedy festival because, as everyone knows, Kilkenny also has a long tradition of economics and humour that stretches back all the way to ... er ... 2010. And, not for the first time, it worked, and Kilkenomics is now one of the most internationally-acclaimed economics and comedy festivals.

4. THE MUSIC AND ARTS AFICIONADOS: With a roots festival and a gospel festival bookending the summer, as well as an arts festival that ushers in autumn, this cohort of visitors make up one of the largest segments of the non-penis hat-wearing tourists that visit Kilkenny.

5. THE HiSTORY CONNOiSSEURS: And last but not least, Kilkenny welcomes those who are drawn to its history and who come to visit what was briefly the medieval capital of Ireland. Along its recently christened Medieval Mile, tourists can stroll through the city's historic centre, which stretches from Kilkenny Castle to St Canice's Cathedral, taking in such sights as Rothe House, the Black Abbey, Grace's Old Castle, Tholsel Town Hall, St John's Priory, the Butter Slip, Talbot Tower, Kyteler's Inn and finally the Smithwicks Experience. This last stop is home to one of Ireland's most distinguished ales, which caused our pubs to smell of old man's farts a full 49 years before Guinness starting doing the same. And let them never forget that!

KYTELER'S INN

You can't come to Kilkenny and not visit one of its fine pubs. And although this is true for almost each and every one of our towns, villages and cities, Kilkenny does have a particularly high concentration of intimate and inviting taverns – and they're not always packed to the rafters with stags. It would be a disservice to all of these pubs to recommend just one based solely on its charm, character and how it makes you feel like you've just met friends you didn't know you had. So I didn't. Instead, I chose one based on the fact that it's the site of the world's first recorded witch trial, which occurred in 1324. Now that's something to think about as you wait for your pint to settle!

Kyteler's Inn, in the heart of Kilkenny City, is said to have been originally established by Dame Alice le Kyteler who, over the duration of her life, tragically lost all four of her husbands to illness. To add insult to injury, she was then accused of their murders! And because she was an extremely intelligent, highly independent, exceptionally motivated, commercially successful, exceedingly articulate, genuinely interesting woman, there was only one obvious explanation – she must be a witch!

Fortunately for Alice, she guessed which way the wind was blowing and jumped bail before Bishop of Ossory Richard de Ledrede had a chance to pass down his judgement. Unfortunately, this meant her maid, Petronilla de Meath, was burnt at the stake in her stead, something that was definitely not in her original contract.

Thankfully, the alleged poisoning of husbands and summary public executions has not adversely affected trade and today, Kyteler's Inn remains one of Kilkenny's many fine pubs and hostelries.

LAOIS

After years of war and isolation, Laois' pristine environment, intact indigenous culture and chilled-out people
mean the county is fast earning cult status among backpackers. Though it is developing quickly, it still retains
much of the tradition that has disappeared elsewhere in
the region. Village life is refreshingly laid-back and even its
capital is characterised by its relaxed riverfront life.

Backpackers come to Laois to experience the rolling mountains, remote villages and tribal crafts. They leave with so
much more: unforgettable memories of smiling locals,
saffron-robed monks receiving alms and markets full of
fresh produce. For those searching for authentic south-east
Asia, Laois doesn't disappoint.

Oops ... sorry ... was mixing it up with Laos. My bad!

ROCK OF DUNAMASE

Over the years, in the classrooms of Laois from Emo to Abbeyleix, Borris-in-Ossory to Ballyfin, it was not unusual that, in the retelling of county history, the O'Mores of Laois were spoken of as a family who had never been defeated by the English. Instead, they had successfully negotiated a peace. In doing so, this put them in the same order of unconquered peoples as the Ethiopians, the Maoris and the Mohicans.

In tales of the O'Mores, the Rock of Dunamase is legendary. Home to their castle, this imposing rocky outcrop, some 45 metres above the flat plain below, proved an impenetrable fortress against their would-be foes and it was from here that the O'Mores led their fierce resistance to English rule. Having constantly terrorised Crown lands and harassed any English forces that strayed into their territory, they eventually forced the King of England into the policy of surrender and regrant – the O'Mores would voluntarily give up their lands, only for them to be returned immediately with titles and deeds.

Though it would be harsh to consider such history a form of Mountmellick creationism, it does fall flat a little in the factual department. While the Rock of Dunamase was home to the O'Mores, it had previously been home to a host of other Irish families before them, including a stay by resident Irish villain Diarmuid MacMurrough. And though the O'Mores did indeed help force surrender and regrant, the small print of their agreement meant they were now expected to speak English, wear English clothes, follow English laws, pay rent to the English King, remain loyal to the Crown, reject the Roman Catholic Church and support either United or Liverpool – hardly unvanquished!

That said, it would be difficult for any true Laois native to look up off the N80 towards the Rock of Dunamase and not feel a sense of pride as they see their ancestral home (of sorts) looking imperiously back down at them.

ELECTRIC PICNIC

Back in 2004, someone had the bright idea to create an Irish music festival with a difference. They wanted to offer a festival that wasn't just about music, a dirty burger and chips and seventeen-year-olds urinating against tents. They envisioned a festival that was as much about the arts as it was about music, open to all ages and full of 'good intentions'. They imagined an Irish Glastonbury and a wonderful weekend out.

All that was missing was a string of empty fields, where not a lot was happening, to put it in.

And this is where Laois stepped in.

Laois, that wonderful, centrally-located county in Ireland that, as luck had it, had quite a lot of empty fields where not a lot was happening.

And finding a home in Stradbally in the east of the county, Electric Picnic was born. For more than a decade, young and old have washed up here to enjoy one of the best music and arts festivals in Ireland, a festival where you can enjoy the headlining behemoths of the main stage and the afternoon humour of the Comedy Tent, a relaxing Sunday massage in the Body & Soul arena and late-night jams by *'isn't that yer man from that band'* at the Salty Dog, legendary sets by *'I didn't know she was still alive'* in the Electric Arena and mid-morning discussions at Mindfield, *'up-and-coming, I saw them when no one knew them'* new musical outfits in the Cosby Tent and family fun at Soul Kids, all topped off with a quiet sleep in a Yurt or a midnight rave in the woods (if only you could find it).

THE TIMAHOE ROUND TOWER

Round towers traditionally doubled up as defensive structures and bell-houses. While there are thought to be about 65 round towers still surviving throughout the island of Ireland, the Timahoe Round Tower of Laois is of particular interest. Built in the 12th century by the followers of St Mochua and standing 30 metres high, what makes Timahoe especially worth visiting is its doorway. Situated some four metres above ground level and providing perfect protection against rampaging Vikings, the Timahoe doorway is considered to be the most ornately designed Romanesque round tower entrance in all of Ireland.

There is little evidence that the tower was ever attacked by the Viking Ragnar Lothbrok or any of his Nordic neighbours. However, there is a local story about a time when it was breached, back in 1837. In this tale, a young man claimed to have climbed unaided to the very top of the tower using only his hands, feet and keen sense of balance. Knowing no one was there to witness the feat and lacking a GoPro camera or its 19th-century equivalent, he decided instead to leave his waist-coat on top of the round tower as evidence that he had managed to climb the whole way up.

Though this story is well known in the area, what is less well known is what weather and wind conditions prevailed at the time – which were probably just as likely to explain how his waist-coat actually got up there.

PORTLAOISE PRISON

Portlaoise prison is Ireland's oldest and only maximum security lock-up. Of all the noteworthy sights in Laois, this is perhaps the least tourist-friendly, if only due to its lack of a gift-shop and baby-changing facilities on site.

Built in the 1830s, with a wonderful entrance gateway that would be worthy of any Norman castle (a feature prisoners rarely fail to notice), for much of its most recent history, the prison has been infamous for housing convicted members of the IRA. As a child, I remember sitting silently in our car with my siblings as we drove by the prison. Situated menacingly on the eastern fringes of Portlaoise town, such was our fear of it that we were scared stiff that any sounds we might make would disturb the prisoners inside and attract their wrath. This fear probably stemmed from the fact that our parents occasionally threatened to lock us up inside if we didn't stop misbehaving, a warning that was backed up as much by the tone of their voice as it was by our lack of understanding of how the Irish judicial system actually worked.

LONGFORD

In 1911, in Castle Forbes, Longford, a wicker hamper containing eight to twelve grey squirrels, alien to Ireland, was presented by the Duke of Buckingham as a wedding present to one of the daughters of house. The hamper was opened on the lawn after the wedding breakfast '*whereupon the bushy tailed creatures quickly leapt out and scampered off into the woods where they went forth and multiplied*' and have been the bane of red squirrels ever since.

Nearly 100 years later, history seemed to repeat itself when a housing estate, alien to large parts of the midlands, containing eight to twelve four-bedroom reasonably constructed semi-detached houses was released on the outskirts of Ballymahon, Longford. Upon release, they quickly began to spread across the country, becoming known as ghost estates and have remained largely unoccupied ever since.

Thankfully, people from Longford, who regularly top the list of county dwellers who don't give a feck what anyone else thinks of them, have happily moved on from both incidents.

And why wouldn't they, when they've got a wonderful little county that remains largely undiscovered with some of the best fishing spots in the land along the Shannon, the Camlin, Lough Gowna and Lough Ree, some of the best white-water rapids in the country on the Inny, wonderful walks from the Royal Canal Way to the Ballinamuck Bog Loop, history that stretches from Aughnacliffe's 5,000-year-old portal dolmen to the historic ruins of the Cathedral of St Mel in Ardagh, and a home rental market that can't be beaten!

THE ROYAL CANAL

One striking characteristic of the Irish countryside, particularly from the midlands eastward, is our canals. A revolution in Irish transport and something that the British built (i.e. that the British got the Irish to build), our canals have made a welcome return to Irish life, thanks to recent renovation work and, though no longer used to transport cargo, are a great resource for their surrounding communities.

In Longford, the younger but longer of Ireland's two main canals begins. The Royal Canal, begun in 1790 and finished in 1817, runs all the way from the River Shannon on the Longford side through to Dublin. While the spur that brings the canal into Longford has yet to be refurbished (needs to be refurbished/will someday be refurbished/just get on and get it refurbished!), much of rest of it has and it makes for one of the nicest canal walks in the country.

AS WELL AS WALKING, SOME THINGS THE ROYAL CANAL SHOULD BE USED FOR

- Running alongside it

- Cycling alongside it

- Fishing in it

- Looking at the ducks and swans in it

- Watching other elements of nature around it

- Jumping into it (only if you can swim)

- Barging down it

AND SOME THINGS THE ROYAL CANAL SHOULD NOT BE USED FOR

- The disposal of road cones

- The disposal of empty beer cans

- The disposal of suitcases, backpacks, briefcases or any other item of baggage

- Driving your motorcycle down it, no matter how iced over it gets!

CORLEA BOG ROAD

Corlea Bog Road, often referred to as the Corlea Trackway, is an Iron Age road built with oak planks around 147–148 BC. Stretching for about a kilometre, the road was discovered buried two metres under the surface in 1984 during peat harvesting. Made up of 100 toghers, or causeways, that during this period would have allowed traffic to traverse an area of land predominantly made up of bog, quicksand and the type of swampy stuff you hope you never get stuck in, it is the widest Iron Age trackway built by the Celts ever discovered in Europe, putting paid to that popular Iron Age theory *'all Roads lead to Rome'*, unless Rome was located in a bog in south Longford.

While we might never know why these causeways were built, we can guess a number of associated pros and cons.

CORLEA BOG ROAD PROS AND CONS

Pro: The trackway must have more than halved the local commute time into the bog.

Con: Who wants to hurry into a bog?

Pro: There were probably no speed cameras on it.

Con: It must have been hard on the suspension and a hoor to overtake on.

Pro: Made of freshly cut renewable wood, the road when first built must have looked and smelt great.

Con: Made of freshly cut renewable wood, it was never going to last and, under its own body-weight and rising surroundings, was on the way to the bottom of the bog within ten years.

Pro: Because it sank into a watery anaerobic environment, much of it was preserved and an eighteen-metre oak stretch of it is now on display in the welcoming Corlea Trackway Visitor Centre where there is never a queue to see this rare antiquity (in your face, Book of Kells!)

Con: That's because the Corlea Trackway Visitor Centre is also in a bog.

ST MEL'S CATHEDRAL

As well as being Longford's landmark building, St Mel's Cathedral might well be Ireland's best-known cathedral outside of Dublin. Though it is a stunning neo-classical structure that stands prominently out in the Longford skyline, it is not its design that many know about but its destruction, when, in the early hours of Christmas Day 2009, it went up in flames.

Though arson was eventually ruled out, the timing of the blaze has meant that suspicion has continued to lie with Santa Claus and to this day, both Comet and Blitzen are personae non gratae in this midland county capital.

Unfortunately, this wasn't the first mishap that had befallen St Mel's. Only a few years after construction began in 1840, it abruptly ceased, when the small matter of the greatest tragedy ever to befall our island (no, not Thierry Henry's handball, but the Famine) meant that building works ground to a halt, at least for a while.

Despite a delay of more than half a decade, the people of Longford got themselves back on their feet and completed the cathedral by 1856. In much the same manner, half a decade to the day after it had burnt down, St Mel's rose like a phoenix and reopened on Christmas Eve night 2014, retaking its place as Longford town's most recognisable landmark.

BALLINAMUCK

FAMOUS
LAST
STANDS

Battle of Thermopylae, aka Leonidas' Last Stand: This took place in 480 BC, when Leonidas and 300 fellow Spartans covered in baby oil (along with some Thebans and Thespians) held the Persians for three days, allowing the Greek army to retreat safely and fight another day.

The Battle of Little Big Horn, aka Custer's Last Stand: Basically the Native American's consolation goal after their century-long battle and eventual defeat by the US Army, when, in 1876, General George Custer and his arrogance were routed by the combined forces of the Lakota, Northern Cheyenne and Arapaho tribes.

The Last Stand, aka *The Last Stand*: 2013 Arnold Schwarzenegger movie that is surprisingly not as bad as you would expect.

BALLINAMUCK'S FAMOUS LAST STAND

The Battle of Ballinamuck, aka Gunner McGee's Last Stand: Ballinamuck is a small town in Longford, where not only will you find a warm north Longford welcome but also multiple memorials and statues to what is generally considered the 1798 rebellion's last stand.

The reason it was the rebellion's last stand was in large part due to the fact that the French arrived in Mayo two months after most of the 1798 rebellion had been defeated.

Having got off to a good start, defeating the English at the Battle of Castlebar and then establishing the Republic of Connacht, things soon went south. This occurred when French leader General Jean Humbert led his men north towards Ulster to reignite the rebellion but then literally decided to go south instead, ending up in the Longford town of Ballinamuck where they made their last heroic stand.

With typical Gallic bravery, General Humbert and his army fought courageously for about a half-hour at which point they surrendered and headed off back to France. And it was probably for this reason that the battle is not called Humbert's Last Stand but is instead popularly known as Gunner McGee's Last Stand. This is because McGee, who led a group of Irish rebel pikemen, remained to skirmish against an overwhelming force of Englishmen and was the soldier who manned a captured six-pounder cannon that fired the final shots in rebellion against the Crown (before our next failed rebellion some five years later).

LOUTH

Long before border cattle raids were popular, the Wee County of Louth was blazing a trail with the original and the best, the Cattle Raid of Cooley, taking place along its wild and windy Cooley Peninsula. Unfortunately, despite being defended by legendary warrior and decent hurler Cú Chulainn, Louth would lose its legendary stud bull Donn Cuailnge, an event that was the first of several unfortunate historical blows to befall the county.

Over the coming centuries, Louth was one of the first places to be invaded and settled, initially by Vikings and then by Normans. Just when things appeared to be settling down, the town of Drogheda was unlucky enough to play host to an arriving Oliver Cromwell, who promptly laid siege to the place, burning large parts of it and throwing all its shopping trolleys into the Boyne before finally slaughtering hundreds of its residents.

It took a long time for Louth to recover and it wasn't until the late 20th century that Louth entered a Golden Age, as the commercial success of Harp lager coupled with the remarkable international success of the Corrs and domestic success of Drogheda United and Dundalk FC combined to boost the Louth economy and the county's morale.

While this confidence briefly took a bump with the loss of the controversial 2010 Leinster final to Meath, the wonderful Wee County of Louth has firmly re-entered the national consciousness and, boosted by those beautiful mountains of Cooley and inviting harbour of Carlingford, is no longer the second last county people remember when trying to list the 32 counties of Ireland.

ARDEE CASTLE

You might wonder if anyone has ever walked out of O'Gorman's Centra in the centre of Ardee with a pint of milk, two sausage rolls and the newspaper, looked up the street and wondered how on earth did the largest fortified medieval townhouse in Ireland manage to get planning permission?!

Of course, it didn't need planning permission: it was here first. It just doesn't seem that way. This is because, unlike other Irish castles, which sit in splendid isolation, Ardee Castle looks like it has squeezed, uninvited, into a row of terraced houses in much the same way you once photo-bombed your friend's communion photo.

Since then, despite the risk of sticking out like a very regal, six-storey medieval sore thumb, it has tried not to become a nuisance and has instead contributed to the town. As well as becoming quite the tourist attraction, it even doubled up for a while as one of the most majestic-looking district courthouses and local administrative headquarters in the country.

MELLiFONT ABBEY

Mellifont Abbey was once one of the foremost religious abbeys in Ireland. As the first Cistercian abbey in the country, it became the architectural model and inspiration for many others built across the land. Despite having much of its stonework 'recycled' over the years for use in the out-houses and dividing walls of the surrounding area, the site still holds much of its charm, with an impressive octagonal lavabo (used by the monks to wash their hands) perhaps the most interesting feature still standing.

Probably the most impressive feature of the abbey was its sheer size. At its peak, it played host to some 100 monks and 300 lay-brothers, which must have led to some gargantuan games of World Cup and British Bulldog. Alongside their very competitive but rarely documented seven-a-side football competitions, the Cistercians were well known for their aptitude in agriculture, hydraulic engineering and metallurgy, which led them to having some of the strongest table-quiz teams among the religious orders.

Unfortunately for the old abbey, its time came to an end in 1539 when King Henry VIII pulled the plug on all the monasteries under his control. Mellifont Abbey regained brief fame (for 'fame', read 'notoriety') when Hugh O'Neill signed the Treaty of Mellifont here, conceding to the English in the descriptively titled Nine Years War. This was followed by another Irish historical nadir when William of Orange used it as his headquarters the night before he won the Battle of the Boyne in 1690 – but then you can't really blame the abbey for this.

MUiREADACH'S HiGH CROSS

Standing impressively at almost six metres tall, Muireadach's High Cross is widely regarded as the most beautiful specimen of Celtic stonework currently in existence. Situated in the ruined monastic site of Monasterboice, some have gone as far as suggesting that this 9th- or 10th-century high cross could well be Ireland's greatest contribution to European sculpture! While this assertion might upset fans of other great Irish monuments such as the Anna Livia in Dublin or that big rusted metal one in Galway (a bunch of boats, you say?), anyone who has ever taken a trip to Monasterboice to look at Muireadach's High Cross would find it hard to disagree.

While every side of the High Cross is intricately designed, it is its east and west sides that demand most attention. To assist would-be visitors in grasping what's going on, I've developed an amateur's guide to Muireadach's High Cross in my typically well-researched and possibly totally inaccurate fashion.

THE EAST SIDE

Traditionally, the east side of Celtic high crosses tend to focus on the Old Testament and keepin' it real and Muireadach's High Cross is no exception. To help you understand it, face the cross directly and then think of it as a hotel with the following floors.

- Basement: A pair of lions playing with each other.

- Ground floor: Adam giving an apple to Eve, while beside him his son Cain is thumping the head off Abel with what looks to be a mandolin.

- 1st floor: Although this might at first appear to be someone holding a lump of pipe about to hit a masked man who's holding one of those really good-value leaf-blowers you'd find in Lidl, this is in fact a scene from David versus Goliath.

- 2nd floor: Moses giving a lesson to a group of Israelites, using absolutely zero active teaching methodologies.

- 3rd floor: A guy picking the pocket of a guy picking the pocket of a guy picking the pocket of one of the Wise Men who've come to hold the baby Jesus.

ॐ 4th floor: Shrek holding a weighing scale containing the soul of a man as he fends off the Devil below with a shovel (or possibly a spade).

ॐ 5th floor: Not a house-party but Judgement Day, with a large group of people to Jesus' right doing 'Rock the Boat' as one guy plays the flute and another the harp. Meanwhile, on Jesus' left, a guy celebrates scoring a goal against the Devil by sliding into a group of football supporters who are 'Doing the Poznan'.

ॐ The penthouse: Hard to see from the ground but appears to be some sort of domestic dispute, possibly because the husband came home late and maybe a little bit drunk.

THE WEST SiDE

Traditionally, the west side of Celtic high crosses focus on the New Testament and are less craic. This side is a lot easier to understand – basically, it is Jesus going up in an elevator and then being crucified. I won't ruin the ending but it's pretty grim stuff, albeit rendered in the most incredibly ornate detail.

CARLiNGFORD LOUGH

Although Carlingford Lough straddles the counties of Louth and Down, because the town of the same name is in the Wee County, I've decided to place it here.

The lough itself is not actually a lough at all but a glacial fjord, as hinted at in its name 'Carlingford'. However, due to the difficulty local people had spelling 'fjord', it has been known as a lough since at least Victorian times, when its combination of sea, scenery and shelter proved a big hit for those coming by rail from either Belfast or Dublin, which are both roughly the same distance away.

Aside from being renowned for its scenic drives, forest parks, rambling, hillwalking and climbing possibilities, Carlingford Lough is known for its two distinct eco-systems on either shore. On its northern coast, it has extensive mudflats and sand-marshes, which help entice migrating pale-bellied brent geese and proves heavenly for bird-watchers; while on its southern shore, the town of Carlingford's extensive range of pubs and clubs lures large group of stag and hen parties who couldn't find any accommodation in Carrick or Kilkenny and heard that Carlingford was just as good.

MEATH

Ah, Meath! Beautiful Meath! While I would stop short of saying that Meath is so great that it, like Fr Joe Briefly in *Father Ted*, should get two parachutes in case one doesn't work, Meath really is a wonderful county. Of course, being from Meath, maybe I'm just a little bit biased.

Many people might know Meath for such things as:

◆ Incredible megalithic monuments at Newgrange and Loughcrew that are older than Stonehenge and every bit impressive

◆ The ancient seat of the High Kings of Ireland at Tara

◆ The largest Anglo-Norman castle in the country in Trim

◆ The site of one of the most defining Irish battles of modern times, the Battle of the Boyne

◆ The first Irish home of one of the finest illustrated book in the state, the Book of Kells

◆ One of the island's finest natural amphitheatres that holds one hell of a rock concert, Slane.

But there are many more wonderful things in the Royal County that people may not know, such as:

◆ A beach at Bettystown!

◆ The original location of Ireland's own Olympics, the Tailteann Games

◆ More racecourses than any other county, Fairyhouse, Navan, Bellewstown and Laytown

◆ Everyone's favourite holiday park-turned-direct provision provider, Mosney

◆ Europe's largest underground zinc mine

◆ And let's not forget UNESCO-listed Tayto Park! (well, maybe not UNESCO-listed ... yet).

Less than an hour from Dublin, as any of the capital's criminal gangs will tell you, Meath truly is a place worth visiting.

NEWGRANGE

Seven hundred years older than the Pyramids of Giza, 1,000 years older than Stonehenge and 5,170 years older than Jenga, Newgrange is one of Ireland's greatest megalithic sites.

Around 3200 BC, a bunch of people we don't know a huge amount about decided to get rocks from lots of places that weren't at all nearby and transport them to the top of a hill and begin building something we don't really understand. And their plan worked!

Covering an area of over one acre, Newgrange is a large grassy mound, surrounded at its base by nearly 100 kerbstones, some of which are richly decorated with megalithic art. Newgrange's official website describes the site as '*a large kidney-shaped mound*' (because what UNESCO-listed site doesn't want to be compared to an excretory organ?). Inside the mound is an inner passage, which is almost 20 metres high, that leads into a cruciform chamber with a corbelled roof.

What makes Newgrange so incredible is that, despite lacking any modern technology, including the internet or Snapchat, it was built in such a way that at dawn from 19 to 23 December, during the annual winter solstice, the sun's light enters through a roof-box towards the front, reaching the floor of the very centre of the passage. As the sun rises, the beam widens within Newgrange's inner chamber until the whole room is illuminated. This event lasts for just seventeen minutes and after that, the chamber goes back to total darkness!

While most accept that this event, which marks the shortest day of the year and the beginning of a new one, is hugely symbolic, we will never know why exactly Newgrange was built, in much the same way we will never know why they didn't connect the Luas lines in the first place.

TRiM CASTLE

A recently fabricated extract from a 1224 medieval *Property Supplement*, found somewhere in Trim:

CASTLE FOR SALE

Hugh De Lacy & Associates are thrilled to present this spacious two-bedroom castle of circa 30,000 m².

Situated in the fabulously modern medieval town of Trim, the castle, tastefully built over a 30-year period, comes to the market with full vacant possession and is grandly decorated. The property is perfectly situated on raised ground, overlooking a fording-point over the River Boyne. Easily accessible (not more than a day's travel) from the Irish Sea, the castle offers a very comfortable escape from the hustle, bustle and intermittent bloodshed of medieval city life.

Unique for a Norman keep, the main building is of cruciform shape, with 20 corners, and is painted white to scare the bejaysus out of the local Gaelic inhabitants. The building is exceptionally secure, with motte, bailey, thick defensive walls, curtain walls, ramparts, portcullis, killing holes, clockwise-winding stairs with hidden steps, stout double palisade and an external ditch surrounding. It also has its own internal well, should you ever be under siege, and out front there are sharp feck-off wooden poles on which you can stick the bloody heads of your enemies.

The previous owner's sophistication is illustrated in tiles imported from Spain and the fact that several of its features are named in French, such as its oubliette, meaning 'forgotten place', which is its onsite dungeon. The attic was converted a couple of years ago and now houses a wonderful great hall, complete with wall hangings made of animal hides. Though golf has yet to be invented, when it is, you will be a stone's throw away from at least three fine 18-hole golf courses.

The property comes with an east-facing chapel, a south-facing main bedroom and a BER rating of J (you'll have more luck heating northern Siberia). Both bedrooms are en-suite and while the toilets (of which there are two) are literally two rectangular holes in the floor, they do double as walk-in wardrobes over which you can hang your clothes so as to get rid of mites, ticks and fleas.

Evening and weekend viewing times accommodated.

TAYTO PARK

MR TAYTO. PART LEGEND, PART POTATO.

Some say he was the son of a dirt-poor mid-western farming family who was taken under the wing of a copper tycoon on the shores of Lake Superior; others suspect that he was the socially awkward illegitimate son of a Russian count who, having shot his wife's lover in a duel, ran off to join the Freemasons; others still say he was a foundling discovered on the streets of Liverpool and raised by a wealthy upper-class family on the moors of Yorkshire; and then there are those who believe him to be just a young Dubliner who done good.

Whatever the truth, by the mid-1950s, Mr Tayto was already making crisps out of a van with two deep-fat fryers. By the turn of the century, he had developed a crisp brand that was an Irish cultural phenomenon and a multi-million-euro business. However, all that would pale in significance with the creation of Tayto Park.

Like young 'Bugsy' Siegel, the man who looked out over a run-down town in the Nevada desert by the name of Las Vegas and imagined a lit-up strip of casinos, Mr Tayto had a vision. Mr Tayto gazed across the farmlands of south Meath, but instead of casinos, he envisioned a Meath Disneyland and so he set about building his dream. Everyone thought he was mad but not five years after it

was opened in 2010, Tayto Park, complete with play-grounds, mazes, pet farms, factory tours, zip lines, zoo, slides, sky-walks and Ireland's first and Europe's largest wooden rollercoaster, has become one of the nation's top ten tourist attractions.

THE HiLL OF TARA

While it might not be as otherworldly as Angkor Wat, as palatial as Petra or as pointy as the Pyramids, the Hill of Tara is just as important culturally to the country of Ireland as these majestic locations are to theirs. From Neolithic times up to the 12th century, the Hill of Tara was held as a sacred site associated with kingship rituals and was the ceremonial capital of the High Kings of Ireland. However, it wasn't just kings that Tara was home to. Celtic pagan druids also employed Tara as their base. Unsurprisingly, it was here that St Patrick's first came to when he returned to Ireland, setting up the mother of all grudge-matches with the druids, which St Patrick won by TKO.

While Tara might now appear at first like a hill with a bad case of the mumps, closer inspection gives a fascinating insight into life back then. Perhaps the most interesting feature at Tara is its *Lia Fáil*, or Stone of Destiny. Seated on top of the hill's most prominent mound, this stone was said to scream once the would-be High King met a number of challenges and was the spot where he was then conferred with kingship. Today, the *Lia Fáil* no longer screams, which is probably a good thing, but does give a damn fine view over Meath and the surrounding countryside.

OFFALY

45%: The percentage of people who are able to correctly spell Offaly

4%: The percentage of people who are able to correctly spell *Uíbh Fhailí*, Offaly in Irish.

Despite being the most misspelt county in Ireland and despite sometimes being described as a large bog in the middle of Ireland, Offaly (pronounced '*awfully*') has so much more going for it as a county than you might first expect – even though it kind of is a large bog in the middle of Ireland:

> It holds the record for being the hottest place in Ireland this past century, with a sweltering 32.5°C recorded in Boora back in 1976.

> It is home to what was, for 70 years in the 19th century, the world's largest telescope, at Birr Castle.

> It is the location of one of Ireland's few volcanoes, the now dormant Croghan Hill.

> It can lay claim to a world-famous whiskey, Tullamore Dew, and a world-famous descendant, Barack Obama.

> Its monastic site Clonmacnoise was, during the 7th century, the educational centre of western Europe.

> It holds the 'Rock the Boat' world record.

> One of Ireland's first proper murder victims was found here. Named Old Croghan Man, he was found in a bog and is believed to have died from a stab wound to the chest more than 2,000 years ago. A file has yet to be sent to the DPP.

> And finally, it cured cancer. Well, not quite – but John Joly was born here in 1857 and he later went on to develop radiotherapy for the treatment of this illness.

CLONMACNOISE

Clonmacnoise was founded by St Ciarán in 544 AD as a place for humble prayer and reflection. With the help of Diarmait Uí Cerbaill, who would later become the last pagan High King of Ireland, St Ciarán constructed a small wooden church 20 kilometres south of Athlone. Though St Ciarán died of a dose of the plague soon after, more churches were soon added.

Clonmacnoise might have remained a cluster of small chapels were it not for its strategic location on one of the few fording-points along the River Shannon. As a result of this, its population grew from just eight men to almost 2,000 people as it soon became one of Ireland's major centres of religion, learning, craftsmanship and trade with many more churches, crosses, graves and towers being built in and around it. Within a few hundred years, it had become so renowned and was so perfectly located on the Shannon, the medieval motorway of Ireland, that monks and scholars from across Europe travelled here so they could come and study during what was Ireland's Golden Age of Learning.

Of course, as anyone from the country who lives beside a motorway will tell you, such express national routes can also bring with them an increase in crime, something Clonmacnoise was no stranger to. During its main period of growth from the 8th to the 12th century, it was attacked and raided at least 40 times. Robberies became so commonplace that it was rumoured that on at least one occasion the Vikings had to reschedule when they turned up, as the monastery had been double-booked and was in the middle of being pillaged by the Normans.

Despite this, Clonmacnoise continued to grow over those 400 years, with the artisans associated with it producing some of the most beautiful Christian artworks made from metal and stone ever seen in Ireland. From within its sacred walls, the Cross of the Scriptures, the Clonmacnoise Crozier and the Book of the Dun Cow were all created.

Coincidentally enough, 400 years is about the same length of time that Clonmacnoise has been the top primary school tour destination for every 1st and 2nd class in Offaly.

OBAMA PLAZA

First came the Declaration of Independence, which found *'that all men are created equal'*. Then, Abe Lincoln's Gettysburg Address argued that *'government of the people, by the people, for the people, shall not perish from the earth'*. After this, there was Martin Luther King's dream *'that one day'* his nation would *'rise up and live out the true meaning of its creed'*. And later still, Barack Obama promised that *'yes, we can'*. Finally, there came Obama Plaza.

While it might be a little far-fetched to draw a direct line of descent from America's Founding Fathers to a fuel court just off the M7, there is the slightest hint of a connection.

This is because, eight-score and six years ago, Barack Obama's great-grandfather, Falmouth Kearney, left the small Irish parish of Moneygall, County Offaly, for the United States in search of prosperity. Settling into a log-cabin in Ohio, he might not have become the most successful emigrant Irishman to cross the water but he did set off a chain of events that one day would lead the 43rd President of the United States back to his ancestral homeland in Ireland. And it is here, back in Moneygall, that

Barack Obama pulled a pint of Guinness, shook every hand in the village and gave inspiration for what might well be the classiest service stop in Ireland – a place you can ponder the American Dream as you fill your tank full of diesel.*

*That Moneygall's Obama Plaza station is actually within the county boundary of Tipperary and not Offaly is just a small little snag, not unlike Barack Obama saying he is from the United States but actually being born in Kenya. (What? He was born in Hawaii? You mean Donald Trump was wrong?!)

LEAP CASTLE

While there are very few castles today that do not claim to be haunted, very few can outdo Leap Castle in Coolderry, both in terms of quantity of ghosts and in the bloody nature of how they came about.

The castle's earliest ghost comes from its most famous murder. This took place in what has since been appropriately named the Bloody Chapel, when, during a mass being celebrated by Thaddeus O'Carroll, his brother, one-eyed Teige O'Carroll, burst in and drove a sword through his back. As a result of this murder, Thaddeus' spirit has roamed the castle ever since. This murder not only created Leap Castle's first haunting, it resulted in the mass becoming one of Offaly's shortest ever masses, second only to Clara Church's All-Ireland Sunday 'Seventeen-minute Mass of '82'.

Thaddeus wasn't alone for long, not with the O'Carrolls, who killed more people than cholera back then. Next up were the unfortunate McMahons. No shrinking violets themselves, the McMahons were a band of 40 northern mercenaries, who were in the castle celebrating the defeat of a clan who were rivals of the O'Carrolls. However, rather than compensating the McMahons for their time or setting up another type of payment option (like a direct debit or something), the O'Carrolls poisoned them instead, adding another 40 souls to wander the castle for eternity.

While such an act might have made it next to impossible for the O'Carrolls to get a plumber in at short notice, it didn't seem to stop them from having more dinner guests over. Evidence of this came in 1922, when workmen discovered the corpses of so many unsuspecting guests who had been dropped through a secret trapdoor onto wooden spikes below that it supposedly took them three carts to bring them out, making the place the worst *Come Dine With Me* location ever.

Consequently, with more murders than a Cluedo convention, Leap Castle is now so full of ghosts that they have to operate a rota system at Hallowe'en every year to make sure every ghoul gets a turn.

BIRR

As well as sounding like the coldest place in Ireland, the town of Birr has the unenviable title of being Ireland's dullest. However, it is important to note that this sad distinction comes from the level of cloud cover it experiences annually rather than from a lack of craic and history, of which it has plenty.

For example, Birr was home to the infamous Crotty Schism. While this might sound like a type of mountaineering injury that would inhibit any man from ever fathering children, it was in fact one of the only times in the nation's history that priests (two Crotty cousins) broke away from the Catholic Church to create their own religion. While their church never really took off, it later became home to a brewery so it wasn't a total failure.

Of most significance to Birr's history is its castle, with a colourful past that enlivens even the dullest of days. Its main attraction is what at first appears to be some sort of impact-collider secret-super-weapon missile-launcher that a James Bond villain might use to bring the world to its knees. It's not that, thankfully, but is in fact a telescope built by William Parsons, the 3rd Earl of Rosse, in 1845. At 52 metres long, six metres wide and known unofficially as the Leviathan of Parsonstown, it is not just any telescope – for over 70 years, it was the world's largest!

But it wasn't just the 3rd Earl who helped make Birr and its castle so interesting. His wife, Mary, was an accomplished photographer whose darkroom is one of the oldest surviving examples in the world; his eldest son, Laurence, the 4th Earl, ended up measuring the heat of the moon; while his youngest son, Charles, was an avid inventor and built one of the fastest boats of his time. Even their cousin Mary Ward added to the interest of Birr Castle, though this was less to do with her work in microscopy, in which she was skilled, than to her falling out of and getting run over by one of the family's creations, a steam car, thus making her Ireland's first road fatality!

Today, all of these histories, with the exception of that car accident, along with Ireland's oldest wrought-iron bridge and the world's tallest hedge can be viewed and relived at Birr Castle.

WESTMEATH

Westmeathians, Westmeathites and Westmeathese do not like being identified as the county 'West of Meath'. Yes, they do share a title. Yes, they were every bit a part of the ancient medieval kingdom of Meath as their eastern neighbours. And yes, though the ancient province may have disappeared, the diocese that ran along its borders has not, with Mullingar, Westmeath, still housing Meath's diocesan capital and cathedral. But this is where the similarities end. After that, Westmeath is very much its own county.

It is home to a host of beautiful and legendary lakes, has its own unique history from the Hill of Uisneach to the mysteries of Fore and now has its own annual music festival in Body & Soul. And let's not forget the successes of Athlone FC and Mullingar's finest musicians, Niall Horan and Joe Dolan, the midlands equivalent to Frank Sinatra with a little less Vegas but just as much tan.

THE HiLL OF UiSNEACH

Depending on who you talk to, the Hill of Uisneach, situated just six kilometres from Castletown Geoghegan, has played a part in just about every significant Irish political, cultural, religious, mythological, geographical and sporting event of the last 5,000 years.

- At one stage, all of the original five provinces were connected to each other at the Hill of Uisneach and it was seen as the very centre of Ireland.

- For a time, it housed the seat of the Kings of Meath and later the High Kings of Ireland.

- It is here where views on a clear day can take in nearly 20 counties and where the first fires celebrating Bealtaine were once lit.

- On its slope is the Cat Stone, said to resemble either a cat watching a mouse or a cat counting to ten with its eyes closed.

- It is here that the goddess Ériú, after whom the country is named, is said to be buried.

- It was here that the likes of Daniel O'Connell, Padraig Pearse and Éamon de Valera held political rallies to unite Ireland behind them and here James Joyce came for inspiration and ideas.

- This was the spot from where Kevin Moran launched the free kick that culminated with Ray Houghton putting the ball in the back of an English net.

- And it is here, surrounded by the remains of circular enclosures, barrows, cairns, ring forts, two ancient roads and a holy well, that every year a Festival of Fires is held to celebrate this sacred space.

While at least one of these statements might be a little far-fetched (or entirely false altogether), the Hill of Uisneach is unquestionably one of Ireland's most mysterious and historical attractions.

BELVEDERE HOUSE AND THE JEALOUS WALL

Long before Ireland was introduced to the US soap opera that we grew to know and love as *Dallas*, the locals of Mullingar were enjoying their own real-life Westmeath version of this drama that, had television been invented in the 18th century, would probably have been called *Belvedere*.

There were differences. While *Dallas* was set primarily in the cattle-ranch of Southfork, Texas, and chronicled the exploits of wealthy Texas oil millionaires with plots revolving around shady business dealings and dysfunctional family dynamics, the real-life shenanigans of *Belvedere* took place primarily in the palatial Belvedere House, Westmeath, and chronicled the exploits of wealthy Westmeath millionaires with plots revolving around shady business dealings and dysfunctional family dynamics.

At the heart of *Belvedere*'s soap opera was Robert Rochfort, 1st Earl of Belvedere, the 18th-century Westmeath equivalent of J.R. Ewing. The *Belvedere* drama ran for three riveting seasons, summarised below.

SEASON 1

The drama begins in Rochfort's family home, the neighbouring Gaulstown House. Robert comes home to find his wife, Mary, in bed with his younger brother, Arthur, a dashingly good-looking and hugely popular Bobby Ewing-type character. (Actually, he doesn't find his young wife in bed with Arthur at all. Instead, he hears she is visiting Arthur and his wife Sarah quite a lot in his absence, probably due to the fact that she enjoys having human company and her husband is an awful eejit. Robert, however, draws only one conclusion: an affair.)

Not known for his measured actions, Robert confronts Mary and locks her away in Gaulstown House for over 30 years while he moves into Belvedere House. In the season finale, Arthur returns home from England to fight for his and Mary's good names. However, Mary has been forced to falsely confess and Arthur is found guilty of adultery, arrested and locked up in a debtor's prison in Dublin where he dies. No one saw that coming!

SEASON 2

Robert comes home to find that his other brother, George, the Gary Ewing-type brother we tend to forget, has built an extravagant mansion within sight of Belvedere House.

George's Tudenham Park is as lavish and splendid as Belvedere House. Riven by jealousy every time he looks at it, Robert sets out to destroy his brother's home. After several failed ideas, he finally decides to build a three-storey Gothic 'sham ruin' to obscure his view of Tudenham Park, because what better way to stop you thinking about your brother's mansion than a three-story fake building you construct to hide it?

The ruin becomes known as the Jealous Wall and is Ireland's largest folly, still visible in its glorious absurdity to this day. Despite the critical acclaim of *Belvedere*'s second season, interest in the drama slumps due to the lack of scandalous behaviour.

SEASON 3

In order to reclaim market share, season three opens with a half-nude Arthur walking out of a shower he was having in nearby Lough Ennell. It turns out Arthur is not dead – Robert had just dreamt it all! The audience at home aren't ready for this and *Belvedere* is brought to a premature end.

While plans to bring out the complete *Belvedere* box set have stalled, the house, gardens, woodlands and Jealous Wall are open to the public to wander through and relive these almost entirely true, real-life stories.

FORE

Perhaps Westmeath's most legendary feature is also its greatest secret: the magical, mythical Seven Wonders of Fore, all situated in and around the village of the same name.

THE SEVEN WONDERS ARE ...

1.
The monastery built upon the bog

2.
The mill without a race or any identifiable water to power it

3.
The water that flows uphill

4.
The tree that won't burn

5.
The water that doesn't boil

6.

The anchorite (hermit) in a cell

7.

The stone above the door

Admittedly, it is highly unlikely that Fore will ever make it to the top of a UNESCO World Heritage list or attract bus-loads of aging German tourists, who, fearful of political unrest in the Middle East, swap the Great Pyramids of Giza for the Seven Wonders of Fore. After all, the first two wonders sound like examples of bad town planning, while the water that doesn't boil doesn't actually exist anymore and the tree that won't burn also won't grow, having died due to the amount of money people keep plugging into it!

That said, the one thing Fore has going for it over the Hanging Gardens of Babylon and the Colossus at Rhodes is that its Seven Wonders are generally still standing and may even increase in the future, with No.8 being *'how did they ever come up with the idea?'* and No.9, *'why couldn't they think of a few better ones?'*

LOUGH DERRAVARAGH

Reflecting its wonderfully soft name, Lough Derravaragh is one of the Ireland's most beautiful midland lakes. Stretching for nearly ten kilometres upwards and four kilometres across, it looks like a skinny lake version of Italy with as much beauty and charm as its Latin twin.

While the lough is well known for its lakeside walks, angling and boating, it is probably most associated with the Irish myth of the Children of Lir.

The story goes that when Bodb Derg was elected king of the Tuatha Dé Danann, Ancient Ireland's supernatural answer to Charlie's Angels, all of his rivals consented to this except King Lir. Rather than just tell him to suck it up, Bodb Derg decided to set Lir up with his daughter Aoibh. This was a success and together with her, Lir had four children: Fionnuala, Aodh, Fiachra and Conn. Unfortunately, Aoibh soon died in a horrible BBC mispronunciation accident. Bodb Derg, sensing Lir's sadness, decided to arrange a second marriage with his other daughter Aoife and it was from this that things went down-hill.

Aoife, jealous of the affection that Lir showered onto his children, plotted to get rid of them. However, neither she nor her servant could perform the gruesome task of murder so instead she turned the children into swans and let them loose into the neighbouring Lough Derravaragh.

While this was obviously disastrous for the kids (there were very few job opportunities in Ireland at the time for swans), what was even more tragic was that the children had to then spend the next 300 years on Lough Derravaragh, 300 years in the Sea of Moyle, and finally 300 years on the Isle of Inishglora in Mayo before a pagan druid would be able to bless them and break the spell. To add insult to injury, by the time 900 years were up, St Patrick had converted all of Ireland to Christianity and there were no pagan druids left!

As to how the story ends, there are several possible endings to the Children of Lir:

~ Some say the Christian monk Mochua in Mayo took pity on the children and was able to break the spell, thus turning them back into withered old people to live out their final days.

~ Others talk of a failed attack by Lairgean, the King of Connacht, on Mochua's sanctuary to capture the swans. During this raid, a silver chain, which had previously been unremarked upon but now linked them together, broke and the swans turned back to humans and promptly died.

~ Others still tell of how St Patrick met the swans, who told him their story. Bringing them back to his house, they heard Christian bells toll and were turned back to humans, at which point St Patrick baptised them. However, because they were now 900-year-old humans, they died soon after. Good job, St Patrick!

~ Finally, in the Director's Cut, after 900 years in the wilderness, they met that last remaining pagan druid who changed them back to humans, at which point Fionnuala is said to have uttered, '*I've seen things you people wouldn't believe. Attack ships on fire off the shoulder of Orion. I watched C-beams glitter in the dark near the Tannhauser Gate. All those moments will be lost in time ... like tears in rain ... Time to die.*'

WEXFORD

When trying to understand Wexford, it is helpful to focus on the importance of two years in Wexford history: 1798 and 1996.

In 1798, there occurred in Ireland the United Ireland Rebellion. Like almost all Irish rebellions, it was borne out of the discrimination, inequity, prejudice and unfairness that many in Ireland experienced under British colonial rule. And, like almost all Irish rebellions, it failed miserably.

However, in the county of Wexford, the Croppies (as the Irish rebels were known) under the leadership of the likes of Fr John Murphy defeated British forces first at Oulart Hill, then Enniscorthy, and finally in Wexford town. Unfortunately, they were unable to follow up this unprecedented three-in-a-row success, losing their subsequent matches before a final heavy defeat at Vinegar Hill. Wexford's success in the 1798 Rebellion is still a source of great pride in the county and is one of the possible origins for their affectionate nickname, the 'Yellowbellies', said to be derived from the yellow sash worn by the local Croppies.

The other possible origin of this nickname is believed to be the county's hurling jersey, which used to sport a yellow stripe across the stomach area. This leads us nicely to Wexford's other important year, 1996, when, after nearly three decades in hurling wilderness, the county hurlers finally retook the Liam McCarthy Cup.

Part of their success came from their inspirational moustachioed captain Martin Storey and his team of young heroes, and part of it came from their rousing county anthem of the time, 'Dancing at the Crossroads'. To this day, the song is still played after 'Amhrán na bhFiann' at closing time in some of the pubs across the county.

Aside from their annual yield of strawberries and potatoes, the final thing you need to understand about Wexford is that their sports supporters are utterly unwavering in their loyalty and, no matter the result, whether their team is beaten by two points or 20, they remember the immortal words of Fr John Murphy as he was about to be hung and decapitated: *'sure, there's always next year'*.

HOOK LiGHTHOUSE

Stretching longingly out from Ireland's south-east coast, Hook Head is one of the country's most scenic peninsulas with breath-taking views across both sand and sea. Probably the highlight of this headland is Hook Lighthouse, one of the world's oldest lighthouses.

The first lighthouse appeared on Hook Head in the 5th century, when an out-of-work monk, St Dubhán, established a beacon to aid local fishermen at the mouth of what would later become known as Waterford Harbour. Then, in a 12th-century episode of *Grand Designs*, it got a remodelling when the Earl of Pembroke, William Marshall, built a 36-metre high tower in its stead, to safely guide ships up to the newly-built port town of New Ross. Though the lighthouse went over-budget and did not include the wet-room shower, underfloor heating or games room originally envisioned, it proved a huge success as groups of monks acted as the custodians to the light, keeping the fires burning and sailors safe.

This continued to be the case until the 17th century, when, after an over-time and holiday pay dispute, the monks were

replaced by professional lighthouse keepers. At the end of the 18th century, more changes took places when the coal fire was replaced by whale oil, which was in turn replaced by gas, which was later replaced by paraffin before finally being replaced by electricity by the 1970s. The last major change to take place was in 1996, when the lighthouse keepers were replaced by digital robots operated from Dublin.

While today its black and white horizontal stripes seem a little dated (all that it is missing are shoulder pads and a perm for that true 1980s look), the lighthouse remains a wonderfully impressive and iconic feature of the Wexford coastline.

BALLINESKER BEACH AND CURRACLOE STRAND

A little-known fact is that Wexford is home to the only successful liberating land invasion in Irish history. This liberating land invasion was nothing to do with the 1798 rebellion, when more than 16,000 Wexfordites, led by a priest (a profession not usually known for its military prowess), managed to wrestle control of three of the four major county towns before being defeated on a hill named after a table condiment.

No, instead the liberating land invasion took place when Hollywood came to Wexford and filmed the Normandy landings, with Tom Hanks, Matt Damon's dead fictional brother and Triple X storming Ballinesker Beach and Curracloe Strand. In doing so, they gained a foothold in France and begin the slow repulse of Germany's Western Front before finally *Saving Private Ryan.* While D-Day didn't actually take place in Wexford, a win is a win is a win, as any of their county hurlers will tell you.

Situated on Ireland's sunny south-east, the two Blue-Flag*
waving beaches, along with their nearby neighbours of
Courtown, Morriscastle and Rosslare, make up the true
Costa del Sol of the Irish coast and this winning combination of white sand, fresh waves and occasional bursts of
sun is why Wexford beaches are among some of the most
popular in the country.

*The Blue Flag: A certification by the Foundation for Environmental
Education that a beach meets its stringent standards of being kept
entirely clear of soiled nappies, Choc Ice wrappers and half-broken
bottles of Budweiser.

LOFTUS HALL

It is not unknown to see children on the way to Hook Lighthouse cowering in the back of their parents' car. The reason for this is the isolated and imposing building that looms over at them as they pass along the solitary road that connects the mainland through to the edge of the Hook Head Peninsula where the lighthouse lies. That building is Loftus Hall, considered one of the most haunted houses in Ireland.

The story goes that in 1766, the inhabitants of the house, Charles Tottenham Loftus, his wife and his daughter, both named Anne, welcomed a young man into their home. He had recently arrived by a mysterious ship that had docked off Hook Head. This was their first mistake – had they been more familiar with the ratings and referencing system employed by Airbnb, they would have realised that this young man was in fact the Devil! To compound this error, when the young man arrived into their house and took off his shoes, the family failed to notice that, instead of having on a pair of smelly odd socks with a hole in the toe like the rest of us, he was actually sporting cloven feet!

Their final mistake was that they ended up getting into a game of cards with him. And, as we all know, aside from the odd round of Hungry Hungry Hippos, the only game the Devil plays is cards. It was during this card game that the daughter of the house, Anne, who, having grown quite fond of the young man, bent down to collect a fallen card and noticed his feet. There's no record of exactly how she proceeded to bring up this fact, though I would imagine it went something like, *'Excuse me, sir, I couldn't help but notice that you've got cloven feet. Is that hereditary?'*

Upset with having been discovered, the Devil revealed himself and flew straight up through the ceiling. In doing so, he left a hole that is supposedly still visible to this day (despite the fact that the building has since been demolished and entirely rebuilt)! Poor Anne, heartbroken that her beau had literally turned out to be the Devil incarnate, never recovered from the trauma and retired to her favourite haunt, the tapestry room. Here she remained, refusing food and drink, spending her days looking forlornly to the sea awaiting his return. He never came back and she died in 1775, nine years later, which is a record in the 'refusing food and drink' department.

Since then, and despite changing hands on several occasions, supernatural occurrences have become commonplace in Loftus Hall, with Devil-induced poltergeists and the ghost of Anne causing paranormal nuisances, scary sightings and unexplained peaks in the electromagnetic fields – all of which visitors can experience during the house's guided tours.

THE EMIGRANT TRAIL

Between the years 1845 to 1852, the Great Hunger occurred in Ireland. Partly caused by the potato blight, a fungus that destroyed successive potato harvests, and partly caused by the cack-handed policies of the nation that neighbours Ireland but that will remain nameless, Ireland underwent a period of mass starvation, disease and emigration that saw a million people die and a million more leave.

One of the best exhibitions to give an insight into the experiences of those who left is the Emigrant Trail of Wexford, which was partly inspired by a descendant of one of Wexford's emigrant sons, former US president John F. Kennedy.

The trail begins at the Dunbrody Famine Ship, which sits on the quays in New Ross, the very quays JFK's great-grandfather left for America at the height of the Great Famine in 1848. Famine ships were among the worst forms of transportation imaginable back in the mid-19th century.

If they were to appear in a game of Top Trumps for 'mass transport', they would be defeated hands down in every category from 'punctuality' to 'passenger safety', 'leg room' to 'sanitary facilities'. With this mind, the only real complaint you could have about the authenticity of the Dunbrody Famine Ship experience is that not enough people come away with a dose of cholera.

Further along the quays is the Emigrant Flame, lit from the Eternal Flame at JFK's grave at Arlington Cemetery. Burning in remembrance of all those who left Ireland's shores, it stands beside the exact spot where he gave his famous 300-word, three-minute long speech back in 1963.

Following this on the trail is the Kennedy homestead in Dunganstown, which, for five generations, housed the Kennedy clan and was the birthplace of JFK's great-grandfather before a final stop-off at the impossible-to-spell-correctly JFK Memorial Park and Arboretum (a fancy home for trees).

While Ireland continues to experience waves of emigration, Wexford no longer plays a key role in these sad departures, with the only real sea-travels from the county being family day-trips to Fishguard and the odd wedding booze run to Cherbourg from Rosslare Harbour.

WICKLOW

If Ireland were a movie, then Wicklow, Ireland's Garden County, would be its trailer. It is the showroom of the country, with a bit of everything you might like about Ireland in it.

- ❧ Enjoy hiking up the Wicklow summits of Lugnaquilla and the tasty-sounding Sugar Loaf? Then you'll love Donegal's Bluestacks and Kerry's Macgillycuddy's Reeks.

- ❧ Happy hanging out in the cosmopolitan and cultured coastal town of Bray? Then a DART ride will bring you to Dublin, which is surely the place for you.

- ❧ Relaxed as you sun yourself at Brittas Bay? Then a world of sandy seaside strands awaits you up and down the west coast.

- ❧ Engrossed in the medieval history of Glendalough? Then a hundred more ancient structures are waiting to be discovered as you criss-cross the rest of the island.

ᰟ Enjoying the great outdoors and greenery along the Wicklow Way? Wait till you see what the Wild Atlantic Way has to offer.

ᰟ And fascinated by cross-community divisions within similar tribes of people that, despite progress, still seem to get sucked back into their brutal past (we're talking Wicklow GAA here)? Then the Ulster question and its troubled history will absorb you.

Wicklow – take it for a test-drive and if you like what you see, visit the rest of the country.

GLENDALOUGH

The monastic site of Glendalough will always be associated with St Kevin, who founded the settlement in the 6th century. St Kevin was the son of one of the ruling families of Leinster, who went off to 'find himself' with a few friends. He ended up in Glendalough, the valley of the two lakes, where, according to reports, he divided his time between praying, living in a tree, sleeping rough, wearing the skins of animals, hardly eating and talking to the birds and animals.

Whatever about Kevin's personal habits, he certainly had an eye for location. Not only is Glendalough just an hour's commute from Dublin, situated in a glaciated valley carpeted with forest that rolls down to two black and beautiful lakes, it is probably one of the most picturesque places along Ireland's east coast.

In these tranquil and splendid surroundings, over the next six centuries Glendalough flourished as a monastic settlement and St Kevin's following mushroomed, with multiple churches and crosses constructed in the area, along with a large cathedral and what has become Glendalough's signature feature, a 30-metre round tower. Though St Kevin died almost exactly 1,400 years ago, his mark still remains, with several of Glendalough's sites of interest named after him, including St Kevin's Church, St Kevin's Kitchen, St Kevin's Cross, St Kevin's Cell, St Kevin's Bed and St Kevin's Hot Tub all dotted in and around this well-loved Wicklow retreat.

POWERSCOURT HOUSE AND GARDENS

While there are many wonderful old stately homes in Ireland, Powerscourt House and Gardens is often considered a class apart. With its combination of a now-renovated Palladian mansion coupled with its formal gardens, it is understandably one of the nation's favourite estates. With a garden of rolling lawns inspired by 19th-century European fashions, which includes cascading terraces and fine-walled gardens, ornamental lakes and shaded ponds, and tree-lined arbours and well-laid out walks, punctuated with garden pavilions and statues including its most famous, the winged horse, it is no surprise that the *National Geographic* once awarded it third place in its World's Greatest Gardens.

And it is for these reasons that had Jane Austen been born in Ireland instead of England, there is a good chance that every RTÉ period production of her novels from *Sense and Sensibility* to *Pride and Prejudice* (which she'd probably have titled *Smugness and Begrudgery*) would have been filmed in the grounds of Powerscourt House and Gardens.

THE SUGAR LOAF

The Sugar Loaf Mountain is a majestic peak that both welcomes visitors to Wicklow and bids them goodbye as they head for the capital. For kids driving past, it always seems much bigger than it actually is, a sort of Wicklow Mount Everest but with a much cooler name and way cooler facts, which, as they grow older, they realise they've been hoodwinked into believing.

THINGS ALL KIDS ARE TOLD ABOUT THE SUGAR LOAF THAT THEY LATER LEARN TO BE FALSE

◆ It's one of the biggest mountains in Ireland. It's not. In fact, it's not even one of the biggest mountains in Wicklow! It just seems big because there is no other summit near it.

◆ While even the most ambitious child probably suspects that the Sugar Loaf isn't made up of sugar, kids are always led to believe that it fits somewhere on the food pyramid. It doesn't. And as they grow up, they learn that the feckers lied to them again when they discover that it is not even comprised of Devonian granite, like its western cousins, but is made of Cambrian quartzite!

◆ Finally, there are whole generations of kids who have driven by the Sugar Loaf hoping it would erupt and that they would have to outrun its sugary lava flows. How cool would that have been? Unfortunately, it is not now nor has it ever been a volcano! It is just an erosion-resistant metamorphosed sedimentary deposit from the sea which, once people started calling it a volcano, didn't have the guts to tell them the truth.

You would think that discovering all this about one of their heroes would leave a bad taste, but children soon get over it. After all, it's only a mountain. And with age, they all realise that, while it might not be edible or about to explode, it is both welcoming and accessible to almost all visitors and even on the most mediocre of days, it offers some of the finest panoramic views of Wicklow, Dublin and the Irish Sea.

AVOCA

Wicklow is blessed with an abundance of picture-perfect villages nestled into its hills, which make superb sightseeing for those who love such quaint Irish hamlets but don't have the time to make it down to Kerry.

Laragh, Aughrim, Roundwood, Enniskerry, Stratford-on-Slaney, Rathdrum and Avoca are just some of the numerous Garden County villages that, between them, have just the right amount of pubs, restaurants, cafés, gift shops and accommodation options to go with their manicured greens and gardens, and that often run alongside a crystal-clear, cool mountain river.

While any of these locations would rightfully hold its own as the perfect destination for a day trip out of Dublin, only one of these has played host to a much-loved television drama that, at its core, was the simmering story of doomed love and unrequited passion. No, we're not talking about *Glenroe* and the village of Kilcoole but *Ballykissangel* and the parish of Avoca.

During the late 1990s, *Ballykissangel* was a BBC flagship programme that followed the love story of Fr Clifford and local publican Assumpta Fitzgerald. At its peak, it landed ten million viewers, meaning every man, woman and child in Ireland was watching it twice at the same time (that or those figures on Wikipedia might refer to the UK audience).

So popular was the series in many Irish households that the name Assumpta became sexy for the first and only time in Irish history, church attendance rates jumped some 76% across the county and, at least for a time, Avoca was the most popular little village in Wicklow.

ULSTER

ANTRIM

ANTRIM BY NUMBERS

1. Shirt number of Antrim's cuddly former hurling goal keeping great, Niall Patterson.

2. Antrim's position on the list of most populous counties in Ireland, mostly due to the vibrant and pulsating Belfast city.

3. The number of times the county's world-famous Bushmills Whiskey is distilled before it makes it into the bottle.

4. A common shout heard on one of the island's finest golf courses, Royal Portrush, just up from Antrim's popular seaside town of the same name.

5. The number of counties that border Lough Neagh, of which Antrim is just one.

6. The number of sides of the hexagonal columns found at the Giant's Causeway.

7. The number of kingdoms of Westeros in George R.R. Martin's *Game of Thrones*, which has become a hugely successful television series with multiple kingdoms filmed in Antrim.

8. The maximum amount of people allowed at one time on the butterflies-in-your-stomach-inducing Carrick-A-Rede rope-bridge.

9. The number of deep green glens that stretch across the county.

10. What I'd give out of ten to a night out in Belfast city.

11. The number of *Taken* movies that Antrim-born Liam Neeson will probably end up starring in.

12. The number of stops currently along the dramatic Gobbins cliff walk at Islandmagee.

THE GIANT'S CAUSEWAY

The story goes that Irish giant Fionn Mac Cumhaill and Scottish giant Benandonner were having a spat over social media or something when Benandonner challenged Fionn to a fight, to which Fionn agreed.

However, as Benandonner was travelling across from Scotland on the causeway that is supposed to have joined both countries in those days, Fionn realised that Benandonner was a lot bigger than his profile picture had led him to believe and that he was in over his head. Thinking quickly, Fionn climbed into a cradle and persuaded his wife Oonagh to disguise him as a baby. When Benandonner arrived, he was first introduced to Fionn's 'baby son'.

Benandonner, being all brawn and less brain, realised that if this was the size of Fionn's baby, then Fionn must be absolutely massive! Offering his condolences to Oonagh on what obviously must have been a tough pregnancy, he ran off, destroying the causeway behind him.

And that is how the 40,000 interlocking basalt columns, the result of an ancient volcanic eruption, whose strange hexagonal shape have made them one of the island's most popular tourist attractions and one of our few UNESCO-listed sites, became known as the Giant's Causeway.

THE DOAGH LOVESTONE

Situated on top of a rocky perch, looking imperiously out over the Antrim countryside, the Doagh Lovestone, also known as the decidedly less romantic 'Holestone', is a pre-historic monument located some fifteen kilometres north-east of Antrim town. Standing almost one and a half metres high with an eight-centimetre-wide circular hole at its centre, the stone is well known locally for its long association with love.

In ancient times, couples who were to be wed stood facing each other over the Lovestone, before the woman passed her hand through the hole in its centre to grasp the hand of her intended and thus consecrate their betrothal – at least until someone more official could bless it anyway. While there is nothing in local legend to suggest that the man couldn't put his hand through instead, the Lovestone's hole is a mere eight centimetres so there is every likelihood that if he stuck it in, he might not be able to pull it out, as an 1833 story of man who had his appendage* stuck in it for several hours goes to prove.

Though the Doagh Lovestone no longer has any official status and to shake someone's hand through it doesn't automatically mean you are now well on the road to adopting a dog, painting the nursery and opening a joint bank account, it is still as popular today as it once was, with visiting couples of all backgrounds coming here to share and shake on their love for one another.

* Appendage, in this case, being his hand.

CARRiCK-A-REDE
ROPE-BRiDGE

First erected in 1755, this 20-metre-long, 30-metre-high rope-bridge was originally built by salmon-fisherman. They did this to join the island of Carrick-A-Rede to the mainland so as to make it easier to cross back and forth. The bridge has outlasted the fishermen and, indeed, the fish, who are no longer caught in any great number off these shores.

Having been renovated and refitted over the years, the bridge has since evolved from what used to look like a set location for an Indiana Jones movie to a modern trust-worthy cliff-crossing open twelve months a year which attracts more than 200,000 visitors annually to what is one of the most northerly parts of this island.

At the time of writing, the fee for the crossing was just under £6 return. What I failed to discover was whether it is possible to get a partial refund on this if you end up among the small minority of people every year whose nerve leaves them on their first crossing and who have to be ferried back from the island by boat!

THE MURALS OF BELFAST

The murals of Belfast have to be among the most famous and infamous works of street art in the world. Tracing the ups and down of Northern Ireland's Troubles and found mostly across east and west Belfast, these gable-end frescoes offer an emotive insight into the psyche of the opposing sides during one of Ireland's most harrowing eras.

Today, the murals are very much a tourist attraction. They can be best viewed on one of Belfast's now famous Black Cab Tours. And as you listen to the colourful descriptions your driver provides while you criss-cross the city looking at these notorious art-pieces, it's worth considering what it must have been like for an estate agent trying to sell and end-of-terrace building back then: '*Oh, lovely location. Very tight-knit community. Great neighbours. Two minutes from the shop. And as for the house, very well-kept. Master bedroom with a bathroom en-suite. Veranda extending out the back, a tastefully restored sitting room and oh yeah ... a 30-foot Red Hand of Ulster/*Tiocfaidh Ár Lá *painted on the exterior wall!*'

ARMAGH

In an episode of *County Family Fortunes*, I asked 100 people to name something they associated with Armagh.

★ Apples: 47

★ Cathedrals: 27

★ Observatories: 20

★ Irish legends: 8

★ Colourful border markets: 3

★ Volcanoes: 2

★ Detailed cattle records: 1

★ Creative fuel practices: 1

★ Hugely successful GAA clubs that used to have a helicopter pad for visiting teams: 1

One of Ulster's border county, Armagh has often been considered a bit of the Wild West of Irish counties, something that stemmed from its difficult past in our Troubling Times, an era of Irish history that is finally receding into the past.

Nowadays, Armagh, the smallest county in Ulster, is much more inviting with no scarcity of natural, historic, cultural or religious sights to see. Of all its characteristics, perhaps its most famous is of the fruit variety: its apple orchards. The rows of trees concentrated in the north of the county, which bloom in May and are ripened by autumn, give Armagh its nickname, the Orchard County, and are the prime reason why, come that week of sunshine in July, we are never short of a cold glass of cider or Cidona to sip on.

ARMAGH PLANETARIUM

When I was young, going on a school tour to the Armagh Planetarium to unlock the secrets of the universe was like buying a ticket to outer space and getting another free. On the one hand, we got to go to the island's premier astronomical theatre where we could go and visit other planets. And on the other, we got to cross the border into Northern Ireland to go and visit another planet – or at least that's what it felt like to us thirteen-year-olds.

With different pictures and shapes on their money, different signs and numbers on their roads and different letters and colours on the registration plates, Northern Ireland was as alien to us down south as Andromeda or Ursa Minor.

Crossing into Northern Ireland is much less of a big deal these days and buses of school kids no longer have to wait at the border as cars ahead are stopped and checked. Instead, they can make it to the Planetarium and be traversing Orion's Belt or the Hoth asteroid field in almost the same time it takes them to finish their maths homework.

SLiEVE GULLiON

Ireland has never had a long and distinguished history of natural disaster movies. Part of the reason for this is that 'our' Hollywood (County Wicklow) is a very different place to 'their' Hollywood (County California). The other reason is because we don't have a long and distinguished history of natural disasters, with our worst ever calamity being the not-very-calamitous-sounding Night of the Big Wind back in 1839.

And because we are lucky not to have devastating hurricanes, tornadoes, earthquakes or avalanches, we are starved of natural-disaster-movie inspiration. Well, almost – we do have volcanoes. While it might come as a surprise to some, there are several of these lava-spewing monsters dotted around the country, waiting for the right type of disaster epic to rekindle a romance. Of my two favourites, Croghan Hill in County Offaly and Slieve Gullion in County Armagh, it is the latter that sounds the most blockbustery. Indeed, you could almost see the tag-line outside your local cinema: *Slieve Gullion – The Giant's Lair*, a name inspired by the hugely impressive children's forest park of the same name in among the scenic forest and heather-covered grounds around the volcano, which our protagonist would most likely be tasked with saving.

That Slieve Gullion hasn't been an active volcano for almost 60 million years is a fine detail that a leap of imagination wouldn't fix. And Slieve Gullion is no stranger to that. After all, it is here in its wonderful walking surrounds that legendary hero and Ireland's mythological version of Ryan Gosling, Fionn Mac Cumhaill, took a skinny-dip in a nearby lake only to emerge with grey hair – though still looking kind of hot. And it was here on Slieve Gullion's scenic slopes that Setanta, arriving late at the blacksmith's Culann's house-party, ended up killing his guard-dog in self-defence and made amends by doing a kind of JobBridge where he worked as a replacement for free until he trained another, thus acquiring the name Cú Chulainn or Hound of Culann.

NAVAN CENTRE AND FORT

Not to be confused with Navan Shopping Centre in County Meath, Navan Centre and Fort in Armagh is one of Ireland's most famous archaeological sites, where myth and reality meet.

Located in the heart of the county, Navan Centre and Fort was one of the great royal sites of pre-Christian Gaelic Ireland and the capital of the ancient province of Ulaid. The history of Navan Fort is a veritable Who's Who of Irish myths and legends with the likes of Cú Chulainn, King Conchobar MacNessa, Queen Mebh, Deirdre of the Sorrows and teenage heart-throbs the Red Branch Warriors all hanging around the fort at various points in its history.

Nowadays, an interpretive centre on the site of Navan Centre and Fort offers visitors a unique interpretation of this history with exhibitions packed full of facts and fables, artefacts and activities, as well as recreations of many of the dwellings of the time along with demonstrations of pre-Christian weaving, cooking and farming. So much effort has gone in to recreate the exact ancient experience, the only real criticism you could make is that the Fort doesn't have the severed heads of enemies hanging from its walls.

JONESBOROUGH MARKET

Situated just over the border from County Louth, although Jonesborough Market serves the general population both north and south, it is probably to those living in the Republic that it holds a sort of cult status. This is because, long before we had our Lidls and our Aldis, Jonesborough Market was the best place to procure consumer items much cheaper than you could find anywhere in the southern counties. That the goods more often than not came without a receipt or a warranty that lasted longer than the time it took you to get out of this small Armagh village is immaterial – they were great value for money. And so for everything from power tools to kitchen appliances, liquors to liquorice all-sorts, knock-down clothes to knock-off CDs, Jonesborough Market had it all.

There was one time of the year more than any other that Jonesborough's already celebrated status rose even higher: Hallowe'en. During this festive time, fireworks that were illegal down south appeared en-masse in the market, ready to set sail to every school-yard in the 26 counties. From bangers to black cats, screamers to spinners, tomahawks to atomic bombs, all sorts of rockets were on hand for an uncle of someone to pick up, bring home and then disseminate to friends, but not before a hefty mark-up to make up for the danger that said uncle had to undergo to safely bring this contraband back home!

CAVAN

Cavan is the type of place that could inspire or incite you, depending on whether you are a glass-half-full or glass-half-empty sort of person.

~ You could grieve that it doesn't have the spectacular natural vistas of the Corks, Clares and Kerrys of this world – or you could just love the fact that it has wonderful rolling drumlin hills and a lake for every day of the year (365 of them, they say) that provide some of the finest fishing in the land.

~ Half of us could bemoan that it has the second highest male-to-female ratio in the country – or half of us could celebrate being spoilt for choice in Ballyjamesduff at the weekend.

~ You could lament its poorly drained clay soils – or you could commend it for being the county best placed to withstand a foreign tank invasion.

~ And while it might have a distinctive scent every so often, as one of the country's largest pig producers, only for Cavan parts of the nation would surely run dry of puddings, rashers and sausages by the weekend.

~ And finally, though its history might not be as well-known as Dublin, Donegal or Derry, with artefacts originating in the county such as the Ralaghan Man and three-faced cult icon the Corleck Head it is every bit as intriguing.

Cavan is best summed up by the local adage, '*if heaven existed on earth, it would be Cavan*' – which contrasts strongly with that other local adage, '*come for the cheaper house prices, stay because of the negative equity*'. The truth, as they say, is probably somewhere in between.

THE SHANNON POT

Splitting the country between east and west, the Shannon is the queen of Ireland's waterways and a treasure for those lucky enough to sail up and down it. While it may not be as long as the Nile, as wide as the Amazon or have as many traffic cones in it as the Lee, it sure is beautiful and is one of Cavan's greatest creations, rising in the north-west of the county in a small pool known as the Shannon Pot.

Coming in at a length of 360 kilometres, which (not that we are counting) is four kilometres more than Britain's longest river (C'MON IRELAND!), the Shannon is not only Ireland's lengthiest waterway and largest river by flow, it also drains a fifth of the country and runs along or through eleven counties.

When not bursting its banks, becoming embroiled in a fraught custody battle with Dublin or falling out of the top ten girls' names in the US, the Shannon lives out its happy existence as the nation's favourite river, offering any number of aquatic activities and drawing a line between the part of the country that's up itself and the part that's not (just kidding, west Athlone!).

And to think it all started from the Shannon Pot. They grow up so quick!

THE CAVAN BURREN

The Cavan Burren is a limestone plateau on the slopes of the Cuilcagh Mountains of west Cavan. Not to be mixed up with its distant relative the Clare Burren, the Cavan Burren is probably Ireland's finest relict landscape or, put more simply, our best outdoor rock museum.

Its stony attractions range from embedded fossils of a 350 million-year-old tropical sea and glacial erratics (enormous boulders) left behind during the last Ice Age to the megalithic wedge tombs of the Bronze Age and 19th-century shelter walls. Basically, if it involved rocks and it happened in Ireland, you'll probably find an example of it here.

While many of these sights might escape the average untrained eye, one of the most interesting features, for the amateurs among us, is the Calf House Dolmen. Part-megalithic monument, part-farm animal shelter, the Calf

House Dolmen is what looks like it would have been a hugely impressive portal tomb, had it been finished. Instead, either due to an accident or an industrial dispute, the final wall is missing and the roof has been left leaning up from the ground.

Though seemingly incomplete as a portal tomb, its design did provide the opportunity for a local farmer, many centuries later, to use its angle to throw a few blocks up inside it to make a very inviting shelter for farm animals. So while it might not now be visually as inspiring as its Clare Burren cousin, the Poulnabrone Dolmen, it is every bit as cute.

LEGEELAN SWEATHOUSE

---- Think sauna. ----

---- Think north-west Cavan. ----

---- Think naked peasants sweating. ----

Think of all these things together and you start to get a good sense of what Legeelan Sweathouse would have been like some 200 years ago.

Despite sounding like a dodgy clothes manufacturer staffed by toddlers, Legeelan Sweathouse, out in the sticks, several kilometres from the village of Blacklion, is one of dozens of long-abandoned sweathouses that are still dotted around the countryside of Cavan. Though these small stone structures can be found, with difficulty, across the island of Ireland, from Rathlin in Antrim to Wicklow down south, the vast majority of their remains are now concentrated in and around the borderlands of Cavan, Fermanagh and Leitrim.

Built with local stone and using a corbelled roof, sweat-houses were often constructed against a mound of earth. Sometimes rectangular, sometimes square, sometimes oval and sometimes all-over-the-shop, they were always small and usually tucked away in or outside rural communities, hidden from the prying eyes of land agents.

Due to the lack of written history around sweathouses, there is some debate as to what they were used for, although we can probably rule them out as forerunners to the men's shed movement, diesel laundering units or holiday homes for Dublin's middle-class. Instead, the most plausible theories are that they were either communal spaces for the taking of hallucinogenic mushrooms or an early type of chemist.

Whatever their use, sweathouses seem to have operated in much the same way across the country. First, the interior was heated, usually by burning turf, wood or heather, before small groups of people entered to sweat it out, occasionally leaving the sweathouse to plunge into a nearby stream before returning back to sweat some more. The whole process took a few hours if the participants were there to try and cure any of the wide range of illnesses or ailments they may have had. If they had taken magic mush-rooms, they could be in there a lot longer.

The decline of sweathouses is thought to have occurred from the mid-19th century onwards with the advent of modern medicine, the late-night pharmacy and three-month gym memberships with access to sauna and jacuzzi for just €100! By the end of the 20th century, the sweat-houses had all but disappeared with the only place in the county where groups of half-naked people could still be observed sweating together being the nightclubs of Virginia and Bailieborough.

LOUGH OUGHTER

With a reputed 365 lakes – one for every day of the year – Cavan is known as the Lake County for good reason.

While every year, there is a whole flooded-inland-drumlin-loving, great-crested-glebe-bird-watching sub-culture of visitor that comes to stay in Cavan (between five and ten of them anyway), a much bigger cohort of tourists are those who come to fish. With a selection of lake dwellers that sound like types of muscle strain but are in fact fish, such as tench, roach, bream, perch and pike as well as the famed Cavan brown trout, you can understand why the county has become a fisherman's dream.

Of all its lakes, perhaps its most celebrated is Lough Oughter. This complex of lakes in the centre of Cavan is often considered the location of some of Ireland's best coarse fishing*.

*When anglers swear profusely during cast-off.

DERRY

Derry was traditionally the Russian Roulette of travel destinations for the out-of-town visitor. You could never be sure if you should ask for a bus ticket to Derry (yayyyy!) or to Londonderry (boooo!), with each choice having a 50/50 chance of upsetting the bus conductor.

Down south, of course, we always knew it was Derry, and never felt a pinch of remorse any time we scribbled out the '*London-*' in front of it in any British-produced travel brochures, maps or atlases that we came across as kids. Even librarians in the Republic were in on it.

This difficulty was eventually remedied a few years back when the city was sensitively rebranded as 'the City of Derry'. Such a diplomatic decision now puts tourists at ease (even if the strong Derry accent doesn't), leaving them free to enjoy the attractions, both rural and urban, of the county, from its fine coastal beaches at Portstewart and the wild Sperrin Mountains to the historic settlements at Mount Sandal and the more modern history of Derry's city walls.

THE
WALLS
OF DERRY

Four metres high, over ten metres wide and running approximately one and a half kilometres long, the Walls of Derry make Derry not only the sole remaining intact walled city on the island of Ireland but also one of the finest examples of a fortified town in Europe.

With a distinctive central diamond at its centre, the city was the first planned city in Ireland and was built between 1613 and 1619 by The Honourable The Irish Society (no, that's not a misprint – that's what they actually called themselves). Its distinctive city walls were constructed to help defend the city from Irish insurgents who opposed the plantation.

So formidable were these defences that they managed to withstand several sieges, including one in 1689 that lasted for more than 100 days, earning Derry its nickname, the Maiden City. This nickname is not to be mixed up with its other nickname, Stroke City, which originates not from the town's proud history of heavy petting but due to the political correctness of calling the place Derry-stroke-Londonderry for much of its troubled history. Fortunately, with the Troubles falling away into the past, the Walls of Derry, once closed to the public for fear that they were an ideal vantage point over the city for more than just tourists, have long since reopened. Today, they constitute Derry's most-visited site and one of Northern Ireland's favourite short walks.

If you like the Walls of Derry, here are some other walls in and outside Ireland you might also enjoy:

— Hadrian's Wall: Pretty ineffective wall spanning northern England that failed to keep out the Scots.

— The Walls: Irish rock band, coming soon to play a local summer festival near you.

— *The Wall Street Journal*: Business newspaper that has everything you need to know about stocks and shares but lacks a good horoscope and the TV listings.

— Great Wall of China: Hugely impressive Chinese construction that used to be known as the only man-made object visible from space, until everyone realised that it was only as wide as the length of their semi-detached house, which is definitely not visible from space.

— *The Wall*: 11th studio album from Pink Floyd. A classic.

— The Walls of Limerick: Traditional Irish dance that's a particular favourite of aunts and visiting tourists.

— The Wailing Wall: Fun-loving peaceful place in the middle of Jerusalem. Very hard to miss.

— The Berlin Wall: Used to separate the part of Berlin where you could find McDonald's and the part where you couldn't.

— *Wall-E*: Hugely successful and critically-acclaimed science-fiction robot-love-story animation movie. What's not to like?

LOUGH NEAGH

While there are five counties in Ulster that border Lough Neagh, there is only one county through which it flows out into the sea and that is Derry. So it makes as good a sense as any to include the biggest lake in Ireland (or Britain, for that matter) as one of Derry's must-see sights.

While Lough Neagh is popular for both its scenery and seclusion, wildlife and water activities (except when the wind gets up) and is home to the most celebrated eels in Ireland, what I find most interesting about the place is its origin, of which there are three explanations.

EXPLANATiON 1
(DERiVED FROM THE TALES OF FiONN MAC CUMHAiLL)

Fionn had yet another argument with Scottish rival Benandonner. Chasing Benandonner from Ireland's shores (although you'd have to wonder why doesn't Fionn just stop inviting him over), he picked up a massive clod of earth, which he then threw at him. Although devilishly good-looking, Fionn had a terrible throw. He missed Benandonner and ended up flinging the earth into the Irish Sea, where it formed the Isle of Man. And because he didn't replace the clod of earth (he'd be such an unpopular golfer), it filled with water and became modern Lough Neagh.

EXPLANATION 2
(DERIVED FROM THE TALES OF EOCHAIDH AND ÉBLIU)

In this tragic love story, Eochaidh fell in love with Ébliu, with whom he had many things in common, including the fact that she was his step-mother! Deciding to elope, they ended up settling at a site where their horse had stopped to urinate (long story). This toilet stop became a spring, which they then covered with a capstone so it wouldn't flood. While today they never would have got planning permission to build within 100 metres of a watercourse, back then things were a little bit laxer and this led to their downfall. One evening, Eochaidh forgot to put the capstone back on the spring when he was locking up for the night and the place flooded, the tragic lovers drowned and the place became known as Lough Neagh.

EXPLANATION 3
(DERIVED FROM THE TALES OF SCIENCE)

Fault lines on the earth's crust filled with water.

MOUNT SANDEL

Ireland's weather is mild, moist and changeable, with abundant rain and occasional sunshine. As a result, it probably never topped the list of 'Most Popular Places to Hunt and Gather' back in the day.

But all this changed in 7600 BC, when a group of Mesolithic hunter-gatherers moved to the island of Ireland and became our first human settlers. Possibly coming ashore along the eleven golden kilometres of Derry's Benone Strand or onto the sandy crescent-shaped beach that would become the ever-popular resort of Portstewart further east or just straight up the River Bann, a small band of Mesolithic hunter-gathers arrived and eventually settled inland on a high ridge overlooking the river in modern-day Coleraine.

Due to the absence of a Centra nearby, they had to rely solely on their wits and skill. But thankfully, due to the abundance of locally-occurring foods such as pigs, fish, birds, seeds, fruit and nuts, they managed to gain a foothold and the rest is history.

Because their original homes were simple tent-like structures, there is little evidence of their time here 10,000 years ago. In fact, it is only by the reamins of their charred hazelnuts and excavated post holes on the site that we know of them at all. However, the surviving imprint of a Stone Age fort nearby serves as a reminder that it was along this parkland stretch of the River Bann that Ireland's first men, women and children arrived, ten millennia ago. And it was here they must have walked through the surrounding woodlands complaining about the weather, just as we still do to this day.

THE BOGSIDE

The Bogside is a predominantly Catholic neighbourhood of Derry that lies just outside and under the shadow of its city walls. It was here in 1969 that a violent three-day battle between its residents and the local police, the Royal Ulster Constabulary, took place. While the riots were sparked by an annual Apprentice Boys parade in commemoration the Protestant victory of King William of Orange at the Battle of the Boyne, they were underscored by decades of wide-scale discrimination against the Catholic population of Derry. The ensuing three-day Battle of Bogside was widely seen as one of the first major confrontations in what would become known as the Troubles.

I have my own personal memory of the Bogside. I studied science, technical graphics and occasionally geography there. However, it wasn't in the Derry Bogside I did this but in the prefab of my Meath secondary school known by the same name to both student and teacher. Of course, the extension wasn't called the Bogside soley due to the teachers' sporadic use of internment, gerrymandering or rubber bullets but also because it was built during the height of the Derry disturbances.

Fortunately, things have changed on all fronts. The Bogside in Derry has moved into a more peaceful era, something visitors can observe as they tour the intricately-painted gable murals that have remained from the Troubles, the foremost of which is the landmark welcome sign, 'You Are Now Entering Free Derry'. Meanwhile, back in my old school, that extension has long since been torn down and has been replaced by a building that is high on student-led learning and low on baton-charges.

DONEGAL

Donegal is probably the most beautiful county in Ireland – with a disclaimer attached. That disclaimer is the weather.

Early reports say that it was God's first hideaway, where he came to get away from it all after six days creating the earth. The story goes that after an afternoon walking the Fanad Peninsula, where no one recognised him because he didn't play football for the county, he ended up in one of those legendary Donegal lock-ins, a few kilometres from Killybegs. Two things came out of this:

1. The first was his exit sometime around dawn, putting paid to his plans to work Sunday.

2. The second was that it was here he first heard the beautiful soft lilt of the Donegal accent. This is why he still returns for a few days every summer, bringing eternal blue skies, warm breezes and a colour that makes the place one of the most memorable in this or any land. On these days, you understand why Donegal captivates and enthrals in equal measure with landscapes of blanket bogs, sheep-studded pasturelands, rugged peninsulas full of character, stark imposing mountains, deep open valleys and a roughly hewn coastline necklaced with white sandy beaches to rival anywhere for 1,000 kilometres.

Then there are the other days, which fall into two other categories.

1. The 'okay days', when it is wet, windy and damp, though still ruggedly beautiful.

2. The 'bad days'. Maybe because that Donegal girl never returned His call, these are the days when the county, the most north-westerly in Ireland and fronting onto the North Atlantic, is a squally hell where windows, doors and curtains are pulled shut as the wind howls and the lights flicker.

And that's Donegal for you. As they say, '*up here it's different*'.

GLENVEAGH NATIONAL PARK

As a national park, it might not be our largest (the Wicklow Mountains) or smallest (the Burren), our wettest (Connemara) or even our most renowned (Killarney), but Glenveagh National Park is certainly one of our most beautiful.

Set over 170 square kilometres of remarkably wild Irish countryside, Glenveagh is a wonderful place to wander around, trekking any of its umpteen walks from the Derrylahan Nature Trail to the Lough Inshagh Walk, the Lakeside Track to the famed and fabulous Bridle Path to choose from. With as much chance of catching a glimpse of a red deer or a golden eagle as you do of being bombarded by midges as big as your fist, Glenveagh National Park is a real slice of adventure in a 'paradiscally' untamed corner of Ireland. (And yes, I made that word up.)

Perhaps the most awe-inspiring chunk of Glenveagh is where it extends into the ice-carved corrie (hollow) infamously known as the Poisoned Glen. Like a setting for a high fantasy movie, the Glen sits at the foot of a real mountain-lovers' mountain, Mount Errigal. With the eye-catching Old Church of Dunlewey also holding court here, hewn from locally-sourced white marble and blue quartzite, it is no wonder that the whole area is one of the most treasured spots in Donegal.

But why then is such an awe-inspiring location named the Poisoned Glen? Thankfully, it's not the result of a disastrous oil spillage here but is instead attributed to two possible events.

The first is the legendary murder of Balor of the Evil Eye, that ancient one-eyed giant king of Tory, by his grandson, Lugh. Balor's Evil Eye was so destructive it had to be covered by seven curtains and, when revealed, would set the whole land alight, making him the least popular guest at a birthday party ever. Legend has it that during the Battle of Mag Tuired, Lugh threw a spear or a sling or scissors or another of those things your mother is always warning you not to run with, and it hit Balor square in the eye. It killed him, but not before he first spun round in pain, setting fire to his own army (hate that) before collapsing onto the ground, Evil Eye still open, splitting a rock and poisoning the glen forever! Forever! Foreverrrr!!!

Then there is the second story, that locals wanted to call the place *An Gleann Neamhe*, meaning 'The Heavenly Glen', but the English cartographer in charge of the process replaced the 'ea' with an 'i', *An Gleann Nimhe*, meaning 'The Poisoned Glen'. Twat.

SLiEVE LEAGUE

The Slieve League

AKA

Slieve Leag

AKA

Slieve Liag

AKA

Sliabh Liag

AKA

Carlos the Jackal

While the last title I can't confirm, the Slieve League sea cliffs seem to have more names than an international jewel thief. I guess when you tower 601 metres above the ocean below, you can have as many identities as you like.

When Slieve League isn't planning diamond heists, its sea cliffs can be found along the Donegal coast (where sea cliffs generally hang out) and are a wonderful part of the county's Wild Atlantic Way. This ridiculously beautiful route traces a slow but scenic path along the coastline from Donegal town through to Dungloe, taking you past the picturesque villages of Glenties and Glencolumbkille, before heading much further northwards up and around the peninsulas of Fanad and Inishowen. Driving this route on those rare sunny days that occasionally occur up here, there are few places on this earth as awe-inspiring.

Though the Slieve League cliffs are only the second highest sea cliffs in Ireland, they are arguably the country's most dramatic (unless you're asked this in in Mayo, when of course you'll agree it's the cliffs at Croaghaun, or in Clare, when the Cliffs of Moher and its car-park take the title hands down).

TORY ISLAND

Though only fifteen kilometres north of the Donegal main-land, Tory Island is truly a land apart. As a local youngster there was said to have once written, *'Ireland's a large island off the coast of Tory'*, a statement that pretty much sums up their staunch independence. While not the easiest to get to, a visit here rarely disappoints.

TO HELP PREPARE FOR A TRIP TO TORY, THE FOLLOWING SHORT TRAVEL GUIDE WILL HELP

- When you arrive on Tory Island, you do not meet the King of Tory – the King of Tory meets you.

- Though you do not need to speak the native language, Irish, it is always nice to say *'Slán'* any time you are leaving. You could also try your hand at *'Dia duit'* (hello) and *'Go raibh maith agat'* (thank you), but if you've never been inside an Irish classroom, then you'll probably murder these phrases and it might be best to leave well alone.

- Tory either means 'stone' or 'pirate' and therefore struggled as a tourist destination for its first 4,400 years.

- To fully, fully appreciate Tory's long, beautiful and stunning summer days, you should really endure Tory's long winter. If hell were cold, it wouldn't look much different from a Tory Island winter storm.

- Light and colour are clearer on Tory.

- A Tory islander's blustery day is a mainlander's storm-warning.

- That's not a cuckoo you hear but a corncrake. Listen carefully before it disappears.

- There are probably more artists, poets, musicians and dancers per square metre and per head of population on Tory than anywhere else in the world.

- If you're not great on boats, do not sit up the front of one during the crossing.

- The East Town–West Town rivalry on the island is not like it used to be.

- Every year, at least five dyslexic film buffs return from Tory disappointed that there was no giant wooden horse on the island.

- If a session gets into full swing, it might be a good idea to ring the boss and tell him you're going to be a few days late before you order your next pint.

- It doesn't matter how many left feet you have, when its 'Céili Time', join in. If in doubt, look for the middle-aged woman who most reminds you of your kindest aunt to lead you through the sets.

- If asked, the first settlers on the island were said to be the fierce pirates known as the Formorians who were once ruled by Balor of the Evil Eye. A later settler group of note was an order of monks led by Colmcille in the 6th century. Though both leaders were Sagittariuses, they shared little else in common.

- Today, the islanders are notoriously friendly and welcoming but they also have an infamous Cursing Stone that once shipwrecked a passing British gun-boat so don't f*ck with them!

- I mean it, don't f*ck with them!

BALLYMASTOCKER BAY

Barack Obama. James 'Buster' Douglas. Susan Boyle. The Teletubbies. Every so often, someone or something comes out of the long grass and takes the world by storm.

Ballymastocker Bay did something similar when, a few years ago, it was named the second best beach in the world by the *Observer Magazine* in a travel article that may never have actually existed but which has been widely quoted ever since.

News that a beach in Donegal was only pipped by some strand in the Seychelles sent people across Ireland off in search of their secondary school atlases to first locate Ballymastocker Bay, and then to wonder why they had never heard of it before! Of course, to those from Donegal, Ballymastocker Bay was old hat and they had long recognised it as the *Baywatch* of the north-west, with just a little less sun, heat and Hasselhoff.

Best arrived at from the direction of Rathmullan and down the rally-hardened hairpin bends of Knockalla Hill, the beach extends for miles, all the way to Portsalon, meaning no matter how famous the beach gets, you'll always have space for your bucket and spade.

DOWN

down¹ *prep.* In a downward direction along or through; e.g. '*I saw Lassie going* **down** *through the fields in the direction of the old mines.*'

down² *adj.* Directed or going downwards; e.g. '*So what you're telling me is, Lassie fell* **down** *the abandoned mine-shaft?*'

down³ *adv.* At or towards a relatively low level; e.g. '*since Lassie died in that mine-shaft, my spirits have gone* **down** *...*'

down⁴ *v.* To drink or swallow quickly; e.g. '*... and I've been* **down**ing *whiskey to help me cope ...*'

down⁵ *n.* Fine soft feathers or hairs; e.g. '*... and I miss his, er,* **down** *the most ...*'

down⁶ *n.* (usu in pl) A rounded ridge or hill, esp. in southern England; e.g. '*... and the ways we would wander aimlessly along the* **down**s.*'

Down[7] *n.* A heavy metal supergroup that formed in 1991 in New Orleans, Louisiana; e.g. *'I wish it wasn't Lassie but **Down** that had fallen into that abandoned mine-shaft.'*

Down[8] *n.* A county situated in the north-east of the island of Ireland that stretches from historic east Belfast to the misty Mourne Mountains, from the natural wonderlands at Strangford Lough to the more manicured beauty of Royal County Down Golf Club, with the best footballing slogan in the whole of the GAA, *'Up **Down**!'*.

TITANIC BELFAST

Located in east Belfast, on the Down side of the River Lagan, lies one of Ireland's premier tourist attractions, Titanic Belfast. The museum and exhibition centre lie on the very slipway where the boat was launched, beside the Harland and Wolff shipyard where it was built.

Of course, while the Harland & Wolff shipyard built many, many ships, only one struck an iceberg off the coast of Canada, sank and then went on to have a hugely successful career in Hollywood.

And while Titanic Belfast is also a monument to the city's maritime heritage, it is the *Titanic*, the so-called Unsinkable Ship, which sank in the early morning of 15 April 1912, 600 kilometres south of Newfoundland, that is at the heart of this exhibition. And it is the story of this ship – its construction, its voyage and the tales of those who travelled on her, including the 1,503 who died that fateful night (1,504 if you count Leonardo Di Caprio) – that is told with great detail through much of its 12,000 square metres of informative space.

STRANGFORD LOUGH

Been to Strangford Lough recently?

If so, did you spot anything unusual?

... A lot of sun tans?

... More people wearing shorts than you would have expected for November?

... Shopkeepers not really interested in the radio weather forecast?

... Insects that you last saw on a David Attenborough African wildlife episode?

... Foreign-looking birds that don't speak a word of English?

... Locals appearing very happy with their lot?

... Difficulty explaining the concept of frost?

If you did, then it might have something to do with Strangford Lough's unique subtropical micro-climate.

While you might never have guessed it, what with it being in County Down and all, Strangford Lough is one of the few places on this island that enjoys subtropical weather all year round, due in large part to the protective arm of the Ards Peninsula. While this does not guarantee the place being entirely free from the odd scattered shower, it does generally avoid the worst of any passing storm and is a no-go area for ground frost. And this is why some of the best gardens and stately homes for miles are dotted around it, with the Mount Stewart gardens at its northern corner being one of the best.

While there are plenty of activities and attractions open to visitors in and around Strangford Lough, one particular stand-out sight is Strangford Stone. Standing over ten metres tall in Delamont Country Park on the southern end of the lough, the stone is the loftiest megalithic monument anywhere in Ireland or Britain. But before you wonder why your history book never mentioned it, it's because it was only erected in 1999 by 1,000 young men and women, who worked together to pull this 47-tonne piece of Mourne Mountain granite into place. And though the stone looks just the smallest bit like that game of Snake you used to play on your old Nokia, when you stand beside it looking out across the Costa del Strangford with the Mourne Mountains behind, you can't help but agree it was a fitting way to mark the millennium.

STORMONT ASSEMBLY BUILDING

STORMONT.
STORM-ONT!
STORM-ONT!!!

With a name like this, is it any wonder that the politics of Northern Ireland has for so long been so tempestuous?

For quite some time, Northern Ireland's parliament buildings in east Belfast, known as Stormont, were a source of resentment to the Catholic half of the population in the province and it is easy to see why. Part of this was due to the fact that Catholics had little or no opportunity to participate in the governing of Ulster. And part of this was probably due to the statues of early Unionist leaders that populated its grounds, with James Craig in the main foyer and what at first looks like Mullingar musical great Joe

Dolan rocking the Eldorado but is in fact that other great behemoth of Ulster Unionism, Edward Carson, on the mile-long avenue leading up to it.

Then in 1998, everything changed when, under the Good Friday Agreement, Stormont became the seat to the new Northern Ireland Assembly. And this great building, built in 1921 for a cool £1.7 million and now open to the public, finally became home to a power-sharing government whose members, even if they don't always share Christmas cards, are at least diverse, representative and peaceful.

CASTLE WARD

Castle Ward is an 18th-century castle with 820 acres of landscaped gardens and a 17th-century tower house on its ground that overlooks Strangford Lough. Open to the public, one of the most intriguing aspects of the castle is the differing tastes of its original owners, Lord Bangor and his wife, Lady Ann Bligh. Lord Bangor preferred a classical Palladian style while Lady Ann Bligh was more a Georgian Gothic kind of gal, and these clashing architectural styles are visible from the outside through to the very centre of the house!

What is also very interesting about Castle Ward is its second history, as capital of the North from which the head of House of Stark rules over his people. Located on the 2,000-mile Kingsroad, which links Storm's End in the south to the Wall in the north, it has been in existence in some shape or form for over 8,000 years.

If you have no idea what I am talking about, then you will also have no idea what most of the people on the Castle Ward tour are nattering about as they talk of dragons, white-walkers and how, despite it being May, '*winter is coming*'.

This is because, like several other locations across Ulster, from the Cushendun Caves and Dunluce Castle to Magheramorne quarry and the Sandy Brae, the quaint and well-kept Castle Ward has been used as a set location for the all-conquering and hugely successful HBO series *Game of Thrones*, doubling up as Winterfell, the home of Eddard Stark, Warden of the North.

FERMANAGH

Fermanagh regularly tops the list of Irish counties that people forget when trying to name all 32, and you can kind of understand why. I mean, with a county town, Enniskillen, that people regularly assume is in Tyrone and villages like Tempo, Lurg, Coa, Belcoo and Boho that sound more like spin classes than settlements, it is perhaps little wonder that when you ask people to talk about Fermanagh, many are left scratching their heads.

But those from Fermanagh who truly know the county will talk about it for hours. They will tell you about its great natural beauty, the deep complex of its Marble Arch Caves, its vast network of nature trails on which kingfishers point the way and the vast open vistas offered by the Magho Cliffs down onto the dark and mysterious Lower Lough Erne. And they will tell you that in summer, Lough Erne is in Fermanagh and that in winter, Fermanagh is in Lough Erne, when the rain pours down from its beautifully scenic countryside to swell these loughs already filled full with drumlin isles.

Now, whether you'll be able to understand their strong Ulster lilt as they're telling you all this – well, that's a different issue entirely.

THE BOA ISLAND FIGURES

Found together in Caldragh graveyard on Boa Island in Lough Erne, the Boa Island figures are two of the most enigmatic and remarkable stone figures in Ireland.

Even though it is thought to be a woman, the smaller and more-weather worn statue is often known as the Lustyman, a name that derives not from her night-time antics but from the fact that she was found on the nearby Lustymore Island.

The other figure is the more detailed and far more impressive-looking Boa Island idol. It is a two-sided statue, one-half male and one-half female, clearly identifiable to the amateur art historian by one side's large pointed male appendage!

With thick square torsos, pear-shaped heads, big, tired-looking owl eyes, no necks, hunched shoulders and crossed arms, what makes the Boa Island figures so enigmatic is that we don't really know what in the name of Jaysus they are doing.

So we asked 100 people what they thought and this is what we got.

42: Standing in a wall defending a free kick

26: Embarrassed as they pose nude for a life-drawing

19: Some sort of dance routine

10: Trying to keep their hands warm as they wait for a bus

2: Preparing for their final dive in the men's/women's three-metre springboard

1: Something to do with the god Janus

LOUGH ERNE

During times of war and under threat of invasion, countries often resort to tactics that will delay, confuse and bewilder their would-be invaders. Ireland demonstrated this during World War II. Unsure whether Germany would use the country as a stepping-stone to attack Britain, the first thing we did was to call the whole era 'the Emergency', in an attempt to make the Germans wonder if they had landed into the wrong period of history. Another action that the Irish government undertook was to remove road-signs or twist them around to point the wrong way, occasionally twisting them back to point the right way in a wily double-bluff.

While coming up with confusingly-titled place names does not appear to have been a further government tactic, it would explain the naming of the two connected lakes that make up Lough Erne, where Lower Lough Erne is to the north and is only occasionally called the Upper Lough while Upper Lough Erne is to the south and is only intermittently called the Lower Lough!

In spite of their confusing names, the lakes are considered to be among Ireland's most charming bodies of water, rightly renowned for their picturesque scenery. The lakes are also well known for their rich mythology and folklore, although many of these tales appear to involve some sort of ancient drowning thus precluding them from inclusion in your average bed-time story. But when not demonstrating ancient Ireland's poor adherence to the basic principles of water safety, the two lakes, which contain 154 islands and as many coves and inlets, make a fine escape for anyone looking to get away from it all.

THE MARBLE ARCH CAVES

FAMOUS ARCHES

Arc de Triomphe: Built to honour those who died in the French Revolutionary and Napoleonic Wars, this iconic Paris landmark doubles as a really tricky roundabout.

Golden Arches: Trusty trademark of a fast-food restaurant known for its ingenious marketing campaigns, which have been getting parents to stop the car at against their will for more than two generations.

Archery: Technically not an arch at all but a pastime that really helps take your mind off things.

Marble Arch: Marble-faced triumphal London landmark through which at one point in history only the Royal Family could pass. This has since changed now that it's part of a traffic island.

Arch-enemy: The person who always gets the shift before you at the local disco and asked yer one to the Debs before you.

Gateway Arch: The world's tallest arch, located in St Louis, Missouri, that celebrates the westward expansion of American pioneers and that was most certainly not part-funded by the Sioux.

THE MARBLE ARCH CAVES

Despite not looking really 'archy' and certainly not made of marble, the Marble Arch Caves rarely disappoint those who come to visit these spectacular limestone caves located just 20 kilometres from Enniskillen. At eleven and a half kilometres long, with underground walkways, boat rides and an informative visitor centre, the Marble Arch complex offers a magnificent glimpse of a subterranean world.

THE WHITE ISLAND FIGURES

Yes, more strange figures in Fermanagh! Not even a kilo-metre long nor half a kilometre wide, White Island on Lower Lough Erne might, for all intents and purposes, seem like just another one of the thousands of small islands and islets that fill our bays, lakes and waterways. It could be considered unexceptional in almost every single way – except for the solitary ruined church that lies on its eastern edge, something that in itself is hardly unique. What makes it worth visiting, however, are the seven figures that face you upon entering that church.

Carved out of a large block of stone and appearing to levi-tate off the ground, each figure is different and each one tells its own story. It's just a pity that we don't know really know what the story was. That said, it won't stop us taking a good guess.

- Beginning on the left, you'll find the smallest form, which is generally considered to be a Sheela na gig, that female character every Irish man's mother warns him about. True to form, White Island's Sheela na gig is naked, squatting and appears to be happily displaying her exaggerated vulva. One of the reasons she did this was to warn off death and evil, something we thankfully don't need to do as much of today due to the advent of seatbelts, fire alarms and a low-cholesterol diet.

- To her right is someone who is either a missionary, Christ or a myopic old man pushing a zimmerframe.

- Next door to him is the largest statue, which could be another pilgrim or an abbot, holding a crozier and a small bag that might contain his lunch.

- The figure in the centre is thought to be David. He is holding a scroll and has his hand up to his mouth in what might be a reference to his role as a psalmist or might be because he's forgotten something really important.

- To the right of David is what at first looks like a young man holding a steering wheel. On closer inspection, it appears to be a young pilgrim wrestling some ancient griffins.

- The sixth figure looks to be the same young pilgrim again but he has returned with a sword and shield. The griffins have understandably legged it. The young pilgrim looks a little pissed off about that.

- The last figure has yet to be carved, which may represent an invitation to the viewer to transfer his emotions, thoughts and wishes onto this unadorned shape – or might just in fact be because it has yet to be carved.

If you don't believe me, then best to take one of those rare summer ferries across and figure them out for yourself.

MONAGHAN

It's says something about a county when their most beloved son, poet Patrick Kavanagh, is famous for writing poems about how terrible the county was. But that's Monaghan for you. They have a great sense of humour.

Kavanagh most famous poem 'Stony Grey Soil' is basically the opposite of a Bord Fáilte advert, in which he accuses the county of having 'burgled [his] bank of youth' and stealing 'the laugh from [his] love.

Of course, Patrick Kavanagh never went to the Oasis Niteclub in Carrickmacross. If he had made it to this Studio 54 of the north-east rather than Billy Brennan's barn, maybe he wouldn't have walked 80 kilometres to Dublin to escape the Drumlin County and his poetry would have turned out very differently.

CASTLE LESLIE

Built in 1870 and fashioned in Scottish Baronial style, if that sort of thing interests you, Castle Leslie is not only Monaghan's most celebrated hotel, it is also a great place to have a wedding – unless you are a former Beatle hoping to keep things secret. While the castle has all the luxury you'd expect of a restored building of its age (beautiful interior, fine lodge, pleasant adjoining stables, charming well-kept gardens that flow down to the lake as well as a fun ghost tour), Castle Leslie is just as well known for the lives of the three Leslie siblings who grew up there.

The youngest was Desmond Leslie, a Spitfire pilot in World War II who pioneered the use of electronic music, was one of the early advocates of UFOs and who shot to wider fame in 1962 when he asked a theatre critic to stand up on live television before decking him in front of eleven million BBC viewers in order to defend his wife's honour!

His older sister was Anita Theodosia Moira Rodzianko King or just Anita to her friends. Sandwiched between two brothers, she too joined the World War II effort, became a mechanic and ambulance driver for the Free French army and reputedly spent at least some of the war's end penning letters home from Hitler's old office!

And finally there was the eldest, John 'Jack' Leslie, the 4th Baronet, who let slip that Paul McCartney was getting married here back in 2002, when he told interviewers that *'it's on Tuesday, but it's a secret.'* Decorated for his efforts in helping delay the Germans at Dunkirk, for which he spent the rest of the war as a POW, he was more warmly known locally for his disco days when, well into his eighties, he started hitting the nearby nightclubs of the Oasis, the Hillgrove as well as those further afield in Ibiza, where he could be often spotted dancing it out to the wee hours!

So if the castle appears to have a unique and fascinating old character to it when you visit, at least you know where it comes from.

ST TIERNACH'S PARK, CLONES

In the small town of Clones, west Monaghan, lies the beating heart of Ulster football, St Tiernach's Park. Here, on what often feels like the sultriest day of the summer, the toughest teams from Ulster do battle. They have to be the toughest because, although Ulster might not always be the strongest footballing province, for a long time now it has certainly been the most competitive. And while in more recent times, these Clones Ulster final days have not always demonstrated the most free-flowing football, they have never been anything but engrossing.

But St Tiernach's Park has always been about much more than the football. It's about the atmosphere on those great July summer days, where pubs the length and breadth of the town are bristling with adversaries and friends. It's about the banter between them in the bar – '*wild hey*', '*go on ye blade ye*', '*bout ya, big mon*', '*some day, hey*', '*he'll take some handlin*' – that continues on past throw-in – '*give him a ball of his own*', '*he's only a cub*', '*linesman, ya saw it!*'. It's about the questionable road-side takeaways that the health authority gives an annual pass to and the bars of chocolate, '*three for a pound*'. And then it's about the colours and flags of friendly Ulster parades as both teams line up before throw-in behind the St Michael's Band.

And then, when it's all over and the Anglo-Celt Cup has found a new home, it's about the traffic jams that never fail to form through Clones' winding streets, which allow every controversy from the coin-toss to those two minutes of injury time to be mulled over in depth.

PATRiCK KAVANAGH CENTRE

Though the poet Patrick Kavanagh may have had a love–hate relationship with Monaghan at times, the county in which he grew up never forgot him. So it came as no surprise that, almost 30 years after his death, the Patrick Kavanagh Centre opened to celebrate one of Monaghan's most famous sons.

The centre is located, appropriately, on Inniskeen Road. It is a good choice because it was on this road at around half past eight on a July evening many years ago, Kavanagh arrived for a dance at Billy Brennan's barn (which at time of writing was for sale on Daft!). However, Patrick didn't go in, possibly because he didn't have enough money for the ticket, didn't understand '*the wink-and-elbow language of delight*', knew he was not going to get the shift or because he'd rather stay outside and be '*king and government and nation*' '*of banks and stones and every blooming thing*'. And as a result of that decision, he ended up outside penning one his most famous poems, 'Inishkeen Road: July Evening'.

Through the use of displays, videos, maps, paintings, not to mention several of his poems, the centre provides visitors with a wealth of information on the life and works of Patrick Kavanagh. As a result, it helps to explain how and why he is so celebrated, as any Leaving Cert Honours student who has learnt their notes will tell you, *'as a poet who had the craft and creativity to transform the ordinary and the banal into something of significance'*.

THE OASIS NITECLUB

There was a time in the '80s and '90s when the Oasis Niteclub was the epicentre of the Irish Saturday night out. While other counties and provinces had their behemoths of disco, the Oasis Niteclub seemed to occupy a realm of its own.

New York, New York

Situated just outside the town of Carrickmacross, for a period of Irish disco history, going to the Oasis Niteclub was a pilgrimage that every young man and woman from the borderlands and beyond hoped to accomplish at least once in their 'going-out-out' lifetime, if not every weekend. Every Saturday, buses from Monaghan, Meath, Cavan, Louth, Leitrim and Longford, as well as further afield, would gravitate towards it so their passengers could disco long into the night.

I can still vaguely remember the one time I made the trip. It was a Christmas in Trim and for the only time I can ever recall, a 50-seater coach was outside our local, taking everyone to Monaghan. Today, the trip to Carrickmacross takes an hour but back then it took at least two, especially with stops, and there would be stops, you can be sure.

While there was a myriad of legends that emanated from the Oasis, it is important to separate what was fact from what was fiction.

Fact: It was the largest nightclub in Ireland when it opened.

Fiction: The Oasis was the first place in Ireland to spell it 'Niteclub'.

Fact: It had the longest bar in Ireland.

Fiction: When time was called at one end of the bar, you could still get another round at the far end if you were quick, due to the time difference.

Fact: Local areas had their own particular spots in the nightclub which operated strict no-fly zones for neighbouring towns. Kingscourt, I am reliably informed, was 'left, beside the bar in the open space between the sofas'.

Fiction: At its peak, 42% of marriages taking place in Monaghan and Cavan and 12% of unplanned pregnancies originated here.

Fact: A bus came up from Cork at least once a month.

Fiction: Post-disco chipper vans were making so much money here during the early 1990s that they were ahead of IT and just behind pharmaceuticals in terms of contribution to Ireland's GDP.

Fact: In order to minimise rows in the car-park, Frank Sinatra's 'New York, New York' was played after the national anthem to send people off in a good mood.

Fiction: The Oasis narrowly lost out on the fifth seat in the 1992 General Election for Cavan–Monaghan.

Fact: Meatloaf played here in 1989.

Fiction: Meatloaf's 1990 hit 'I'd Do Anything for Love (But I Won't Do That)' was inspired by him seeing a couple from Castleblaney chewing the face off each other in the Oasis.

Unfortunately, the good times couldn't last forever. By the turn of the millennium, the Oasis had turned its back on cheesy pop tunes, slow sets and the occasional mosh and had become purely dance. Numbers started to fall. Buses stopped coming en-masse. The writing was on the wall and when, during a huge Garda raid in the early 2000s, 43 people were arrested for 'not being' in possession of drugs, its time was up and with that the Oasis finally dried up.

TYRONE

Second largest county in Ulster? Sixth largest county on the island? One of two counties in Ireland to begin with the letter T? Only county in Ireland that begins with three consonants? One of just two counties in Ireland that could possibly be used as the first name for a boy without fear of him getting bullied? Not the most interesting of facts? Who cares?! Tyrone certainly doesn't.

That's because while the county might appear to some as a tough, resolute, no-nonsense, frontiers type of the place (albeit one with no frontiers), 'West of the Bann' Tyrone is one of the most laid-back places in Ireland.

Perhaps no one sums Tyrone up better than one of its favourite sporting sons, golfer Darren Clarke. So universally well-liked is Darren that not only is he the golfer most golf fans would like to go out for a pint with, he is also the sportsperson fans of *other* sports would like to go out for a pint with!

But that's Tyrone people for you. And that's the reason once you get to know someone from Tyrone or '*among the bushes*', as they sometimes say here, they soon become one of the people you'd most like to go for a pint with, with no better place than in one of those fine wee pubs from Ballygawley to Strabane, Dungannon to Castlederg.

THE HiLL OF
THE O'NEiLL

The people of Tyrone are a proud people with a proud history. The county's name is derived from the Irish *Tír Eoghain*, meaning 'land of Eoghan', Eoghan being a descendant of Niall, the legendary ancient Irish king and mercurial frontman of the prog-rock group Niall of the Nine Hostages. Though the county is now the sixth largest in Ireland, it used to cover a much larger territory when, as late as the second half of the 16th century, it stretched well into modern-day Derry, making it one of the island's most powerful kingdoms.

Further pride stems from the county's association with the O'Neill family, whose red hand marks the county flag, and who could probably hold claim to the title of 'Gaelic family that came closest to almost beating the English' during our 800-odd years' colonial rule.

The O'Neills, under their talismanic leader Hugh O'Neill, put Elizabeth I under huge pressure during the Nine Years War and were within touching distance of victory. Indeed, had it not been for the Spanish landing at the wrong end of the country, a 400-kilometre winter march, hot-tempered Red Hugh O'Donnell being unable to follow instructions and Lidls in Ulster not stocking horse-stirrups (don't ask), the O'Neills might have routed the English army down at Kinsale in 1601 and forced the Crown into ceding independence. As it was, the O'Neill forces came runners-up in the battle, accepted defeat and within ten years were trying to learn the Italian for '*excuse me, is there a pub nearby where can I watch the Gaelic?*'

Today, you can still visit their ancient family fortress on the Hill of The O'Neill, which overlooks the town of Dungannon. Having housed the British Army during the Troubles, who had recognised its strategic importance, the hill is now home to the much more welcoming Ranfurly House Arts and Visitor Centre, where the Hill of The O'Neill's history, as well as that of the wider province, is retold.

THE SPERRiN MOUNTAiNS

IDEAS THE IRISH GOVERNMENT CAME UP WITH TO SAVE THE ECONOMY FOLLOWING THE 2007 BANKING CRISIS

~ Ask the 70 million-strong Irish diaspora to buy a Certificate of Irishness for only €45, starting with Tom Cruise.

~ Ask anyone with an Irish connection to come over here for a 'Gathering'.

~ Ask anyone who has a connection to Ireland and who is famous, wealthy or famously wealthy to come to Farmleigh and tell us what we should do.

~ Ask Martin Sheen, Paul O'Connell and Saoirse Ronan to ask us to ask someone we know to set up a company in Ireland.

~ Ask the ECB, IMF, EC, WB, UFC, C3PO and any other large international organisation with initials to come to Dublin and tell us what we should do.

~ Ask the Chinese to buy Westmeath.

~ Mine the whole Sperrin Mountains and sell the proceeds to a Cash for Gold store in Dublin's Ilac Centre.

In the end, although we might have tried everything else on the list, we didn't mine the Sperrins, leaving them as one of the largest upland areas on the island. Of course, the only reason we didn't was the Good Friday Agreement, a devolved power-sharing arrangement that includes the small detail that Dublin doesn't actually govern Tyrone in the first place!

And it was a good thing they weren't mined. Because while there might be *gold in them thar hills*, the bracken-covered, wild, windy, back-end-of-nowhere mountains, which run all the way up into Derry, provide some of the best hiking trails the North has to offer.

THE ULSTER AMERICAN FOLK PARK

In the noughties, you left because the construction sites had dried up, the economy had gone south and half of your Junior A team-mates were already getting tans in Brisbane. You got a lift to the airport in your parent's Octavia, wearing blue jeans and your county jersey, and took a 777 to a new life and a place Down Under or in Dubai. Others took the boat to England.

In the eighties, you left because there was zero employment for your age group and hadn't been for years. You travelled to the airport in your folks' Fiat 131, wearing blue jeans, a denim jacket and big hair, and took a 747 to a new life and an apartment in Boston or the Bronx. Others took the boat to England.

In the sixties, you left because you were the fifth child of eleven kids on a family farm with no long-term prospects. You made it up to the airport in your oul pairs' Triumph, wearing a suit and tie with an occasional mini-skirt or flare, and took a 707 to a new life on a camp in Canada. Others took the boat to England.

Any time before the sixties, you only took the boat.

While I can't do justice to all the stages of emigration our people underwent over the past three centuries, the Ulster American Folk Park just outside of Omagh can. Here, in an open-air museum, those three centuries of Irish emigration are illustrated, with more than 30 exhibit buildings that take the visitor from the thatched cottages of home, on board a full-scale emigrant sailing ship (sans the typhus) to the log cabins of the American Frontier. Led by the actual people who took these journeys (or maybe just well-versed and -costumed actors), the full Irish emigrant experience is retold in authentic detail.

ALTADAVEN WOODS

Altadaven Woods, which lie along the 1,000-kilometre circular walking route that is the Ulster Way, are a fern-filled delight in south Tyrone. Looking like a chunk of tropical forest, except a little damper, cooler and less hot and humid, there are two well-loved attractions within, a giant chair and a small grove, that, depending on who you listen to, have very different origins.

According to those of a pagan bent, the chair is a mighty seat carved out of a great chunk of sandstone, some two metres high, known as the Druid's Throne. If you sit on it and wish really hard, your wish might well come true. The small grove is down some steps from the throne, where you'll find an altar dedicated to the pagan god Brighid, complete with pagan cup-holders. Here, druids used to perform their rites and rituals, with the occasional immolation of some poor victim thrown in.

If you are more inclined to the Christian story, the seat, still two metres high and made of sandstone, is known as

St Patrick's Chair. And instead of wishing while you sit on it, you have to pray really hard for something to come true. And then down the steps from the chair, the small grove is not a pagan altar but rather a holy well dedicated to St Brigid. The cup-holders are in fact urns of water that miraculously never dry up. It is not a site of rites and rituals and the odd human sacrifice but is instead a great spot to cure warts.

Irrespective of whether you think the places are pagan, Christian or just cool natural phenomenon, the sites are still quite special and, positioned deep into the pristine Altadaven Woods, are a welcome woodland mossy retreat far from the madding crowd.

CONNACHT

GALWAY

While we may never know in which county the word 'craic' was first uttered, Galway is as likely a place as any. The reason for this is, as anyone in Ireland will tell you, Galway is *great craic*.

Of course, in some ways, it shouldn't be. After all, Galway is a place where it's often windy more than 365 days a year and where rainfall is measured not in millimetres but in days. Yet there is something about the place, encapsulated in its central city, that makes Galway *fierce craic*.

Maybe it is the city's mix of people, from the old timers chatting all year round as they carefully fold their clothes before jumping off Blackrock diving board to the blow-ins from around the country and further afield, who cross the Shannon for the summer and never leave.

Or maybe it is the city's setting, from its historic Claddagh village to its westside, where a few minutes in any direction you can be among stone walls and wild ponies. Whatever the reasons, Galway city is *brilliant craic*.

Of course, it is not only the city. North of it, you have places such as Corrandulla, Headford and Dunmore that are *savage craic*. East of it, are the villages of Athenry, Monivea and Craughwell that are *some craic*, while west of it, there's Barna, Spiddal and Oughterard where there is *mighty craic* – although the less said of the southern settlements of Gort and Ardrahan, where there can be *mad craic altogether*, the better.

DÚN AONGHASA

🌀 Paranoid that someone will sneak up behind you?

🌀 Interested in a panoramic ocean view?

🌀 Not worried about getting connected to the water supply this millennium?

If the answer to all these questions is a resounding yes, then Dún Aonghasa (or Dun Aengus, as it is known in English) is the place for you.

Constructed 1,000 years before three wise men with at least two crap presents went looking for a stable in Bethlehem, the prehistoric fort that is Dún Aonghasa stands proudly facing the Atlantic at the edge of Inis Mór, the largest of the Aran Islands.

With a 100-metre sheer cliff-drop down to the ocean for a back garden, not only does Dún Aonghasa provide never-ending photographic opportunities for brave tourists, the fort is also effectively impregnable from behind, meaning you could leave the back door on the latch if you decided to tip off to the shops and have no fear of a burglary.

And while it is not the type of place you'd allow Seanín out to play frisbee, growing up in the fort with its vast views over the Atlantic, taking in the Burren and County Clare to your left and the immense Atlantic Ocean to your right, must have put a different spin entirely on the term 'home schooling'.

Today, one of the nicest ways to visit what one 19th-century artist called '*the most magnificent barbaric monument in Europe*' is to rent a bike in the island's main town of Kilronan and cycle the seven kilometres to its interpretive centre from where you can walk the last few hundred metres. Once your visit is finished, with or without the almost obligatory photo of you peering out over the edge, the prevailing wind and gravity should help carry you back down to Kilronan's harbour and straight into one of the several warm and welcoming village bars for a quick pint before you catch the ferry home.

THE STREETS OF GALWAY CITY

Discovered by Christopher Columbus in 1477, Galway is arguably the one city that every other city in Ireland would agree is not shite, meaning the 6th most populous city on the island of Ireland is probably our most popular. One of the nicest part of the city is its pedestrianised central streets, which stretch from Eyre Square down to the Quays.

Streets full of buskers and performers, from the guy with the bagpipes to yer man who used to put his head in a bucket. Streets with a club sandwich up top and a battered cod and chips down below. Streets where a Spanish student with a sign points one way and two old women chatting point another. Streets where there is always time for a can in the sun down at the Spanish Arch or a pint indoors at a snug in Neachtains. Streets that can be sunny one minute and torrential rain the next. Streets that ooze character and radiate charm.

Streets that will always be Galway.

CONNEMARA

Even the name would make you want to take off there looking for love. Except of course you shouldn't. Otherwise, all you'll end up looking for is your way out of it, until Galway Mountain Rescue find you clinging to a wild pony for warmth, barely alive at the edge of a bog.

Legend has it that Oliver Cromwell had originally intended to tell those displaced in his plantation to go '*to Hell or to Connemara*', until one of his lieutenants mentioned that, even for him, this was a little harsh and he should change it to '*to Hell or to Connacht*' to give them at least half a chance.

While Connemara might at times be an austere and harsh land, it is stunningly beautiful when the sun shines, with wonderful rugged mountains from the Maumturks to the Twelve Bens, sparkling black-lakes and bright yellow furze from Roundstone to Clifden, the beautiful Kylemore Abbey, and golden beaches that spread all the way out into the Atlantic. Indeed, on these days, it is truly, truly magnificent.

And on the other 364 days ... well, that's a different story.

THE TUROE STONE

Carved in relief, standing at more than one and a half metres tall and made of granite, the Turoe Stone is probably the best Irish example of the ancient Celtic style of art known as La Tène.

For the unversed, La Tène art is one based on an abstract curvilinear style of decoration. Basically, it is like that really cool set of doodles that the girl you fancied but were afraid to ask out in your college philosophy class used to draw along the side of her refill pad.

The designs of the Turoe Stone resemble the more famous stones of Newgrange, with their array of mystifying motifs, spirals, circles and curves. However, it is important to remember that they are very different, coming from two distinct eras. The only real similarity they have is that the meaning of the designs and purposes of these stones has been lost in the mists of time. That said, the phallic shape of the Turoe Stone has led several experts to postulate, not for the first time, that an ancient piece of art might have links with a fertility cult.

Originally located a few kilometres from Bullaun, the stone was moved in the 19th century to its home at Turoe Farm. Here it enjoyed spells sitting on a fancy-looking cattle grid, before being banished into what, for all intents and purposes, was a garden shed. While officially the stone is now said to be in Athenry, where it is in the process of being cleaned and restored, there have been rumours that it is actually in a Betty Ford-type clinic receiving treatment for an addiction to prescription medicines. While this seems a little implausible, what with it being a stone and all, it would at least help explain why its return to Bullaun continues to be delayed.

Due for return very soon (irrespective of what month or year you are reading this), on the upside, if it's still not back when you arrive on site, you can always visit the hugely entertaining neighbouring Turoe Pet Farm and feed some goats.

LEITRIM

If Ireland is the underdog of western Europe, the province of Connacht is the underdog of Ireland and Leitrim is surely the underdog of it. And we all know that everyone loves an underdog.

With lovely people, when you can find them (it is the most sparsely populated county in Ireland), the River Shannon running through it, at least six kilometres of Atlantic coastline, the second smallest chapel in the world, a beautiful waterfall, the highest rock tower in the country and a county town that has been great craic for at least a generation, the Ridge County has many reasons to be happy (at least seven, anyway).

THE COSTELLO MEMORIAL CHAPEL

Completed in 1653, following two decades of construction, the Taj Mahal is widely considered not only the jewel of Muslim art in India but also the ultimate symbol of love. Commissioned by Mughal emperor Shah Jahan on the southern bank of the River Yamuna in Agra, the Taj was built as a mausoleum for the emperor's favourite wife, Mumtaz Mahal, who died as she gave birth to their fourteenth child.

The same but different is the Costello Memorial Chapel in Carrick-on-Shannon, Leitrim. Built just over two centuries after the Taj, the Costello Memorial Chapel is also a mausoleum built by a grief-stricken husband, Edward Costello, to remember the love of his life, Mary Josephine.

Unlike the Taj Mahal, the Costello Memorial Chapel is not a UNESCO World Heritage Site nor likely to become one anytime soon. Nor did it cost anywhere close to the €750 million it took to build the Taj and it doesn't welcome anywhere near the three million visitors annually who come to be inspired, to propose and to take funny photos in front of the Taj. However, like the Taj, what sets the chapel out from any of your other run-of-the-mills mausoleums is its size.

At four and a half metres by five metres, about the same dimensions as your average garden shed, the Costello Memorial Chapel is the world's second smallest chapel and – if you can find it – a wonderful lasting tribute to love in the heart of Leitrim.

GLENCAR WATERFALL

Where the wandering water gushes
From the hills above Glen-Car,
In pools among the rushes
That scarce could bathe a star ...

– 'The Stolen Child' by W.B. Yeats

Though a story about the abduction of a child could hardly be considered the best promotion for a beautiful natural wonder like Glencar Waterfall, you can get away with it when it's a poem by the nation's favourite writer, W.B. Yeats. In fairness to Yeats, 'The Stolen Child' is a tale set to verse that is less a bundled-into-the-back-of-a-van-using-a-packet-of-Smarties abduction but more the return-to-innocence-by-running-away-with-the-fairies sort, i.e. the best type of childhood abduction.

Thankfully, no one in the tale comes to harm and the poem can be appreciated for its fond references to several scenic locations from Yeats' childhood, such as Rosses Point and Sleuth Wood of Sligo and the impressive fifteen-metre cascade surrounded by verdant foliage that is Leitrim's Glencar Waterfall.

EAGLE'S ROCK

EAGLES ROCK!!!

Common urinal graffiti from the 1970s, celebrating the Los Angeles rock band whose hotel in California was *such a lovely place* that had *plenty of room*, *any time of year*. Often immediately followed by the retort *Eagles Suck!!!*

EAGLE ROCKS

Either a human-interest story at the end of the news about a guitar-playing bird of prey or a tell-tale sign that usually indicates that this bird of prey is about to faint.

EAGLE'S ROCK

At 330 metres tall and situated in north-west Leitrim, this is the highest free-standing rock tower in Ireland. If you manage to climb up onto its heather-covered summit, which very few people have, it's one hell of a place for a picnic. And if you can't, then that's fine too, as Eagle's Rock and its surrounds present a wonderfully unique vista and a truly scenic place to visit. What is more, with the right kind of light and a well-crafted selfie, you could even kid your friends into thinking you're holidaying in Arizona or the Australian Outback.

CARRICK-
ON-SHANNON

Not as coastal as Carlingford or as cosmopolitan as Kilkenny, Carrick has a reputation as the western hub for stags and hens. However, there is more to it than that, as it is equally famous for being the cruising capital of Ireland.

Thankfully, this title has nothing to do with hanging out in its bushes or public parks. Instead, it refers to its prime boating location on the River Shannon, something that has been further enhanced since the Shannon–Erne waterway was developed. While the town's marina will never be St Tropez, Marbella or Capri, it is perfectly placed to offer some of the finest opportunities for those wishing to set sail and take in the ultimate river experience along Ireland's premier waterway.

MAYO

Mayo. Not the high-in-omega-5, made-from-eggs, *'Bring on the Hellman's, Bring on the Best'* spread that tastes good with pretty much anything, but Mayo the county. County Mayo's motto is *'Dia is Muire Linn',* which roughly translates as *'God and Mary with us ... because luck sure as hell isn't'.* While I may have added the last part, there does seem to be something about Mayo that just seems to invite misfortune.

It lays claim to being perhaps the wettest region of Ireland, with the area of land between the Maumturk and Partry Mountains receiving nearly two and a half metres of rain a year. It is pure windy and is half-covered in sub-standard soil and blanket bogs. It tends to get hit hardest when it comes to famines and its football team haven't won an All-Ireland title since they were said to have disrespected a funeral in Foxford, some 60 years ago, and were cursed by the local priest.

Despite these setbacks, Mayo does have a lot going for it. It's wild and wonderfully rugged, with the scenery and people to match. It has some beautiful, warm and welcoming towns, even if some are impossible to hitch-hike out of (I'm looking at you, Westport). Then there is Achill, the largest and one of the most beautiful islands off the coast of Ireland. And finally, if you fancy a quick prayer, then there are no better places to go in the country than up Croagh Patrick or over to Knock.

So before we feel sorry for Mayo's inhabitants, we should probably first feel sorry for those who've never been.

CROAGH PATRICK

People in the know call this the Reek. '*I was just up at the Reek*', they'll say, or '*Thought I'd stretch the legs up the Reek and have a bit of an oul pray*'. To everyone else, it is known as Croagh Patrick, a 764-metre mountain that is the most important pilgrimage site in Ireland.

Every year, tens of thousands of people climb it, with the last Sunday in July, 'Reek Sunday', proving the most popular. On this day, up to 30,000 pilgrims set off for the top by foot, by barefoot or by knee. During my sole ascent, I hiked with a small pebble in my shoe, partly in contrition for my past sins, partly because by halfway I was too tired to take it out.

Croagh Patrick had been a site of pagan pilgrimages since 3000 BC. Then, during the 5th century, it underwent a Gok Wan-style makeover to become a site of Christian pilgrimage, with none other than St Patrick completing a 40-day fast on top to celebrate how fabulous it now looked. Since

then, it has continued to grow in popularity, particularly among tourists, the devout and the aged, as well as large numbers of secondary school students whose annual class tours get caught in its gravitational pull.

Finally, if improving your fitness and admiring the fantastic views over Clew Bay weren't enough to motivate you, if you put in that little extra bit of effort and are able to complete 43 Our Fathers, 43 Hail Marys, three Creeds and a *'whatever you're having yourself'* while climbing, you can then pray to the Pope for an oul favour or two.

THE CÉIDE FIELDS

EXTRACTS FROM A CONVERSATION BETWEEN HUSBAND AND WIFE SOMETIME CIRCA 3500 BC

Husband: (Arrives into cave visibly upset. Grunts loudly in that Neolithic type of way and sits glumly on a rock in the corner.)

Wife: *'What's wrong, love?'*

Husband: (Silence.)

Wife: *'C'mon. You can tell me. It's not good holding onto these things. Just because you are a man and a hunter-gatherer doesn't mean you can't talk about your emotions.'*

Husband: (After some hesitation and a few more Neolithic grunts) *'I lost my job.'*

Wife: *'You what?!'*

Husband: *'Actually, I kind of resigned.'*

Wife: *'What did you do that for?!'*

Husband: *'I've had enough of this hunter-gathering lark. I'm just not cut out for it. Miles of walking just to find an animal. Then hours of ducking and diving, creeping and running, before you have a split-second to kill a rabbit with an axe from 30 yards! It's not easy, you know. And I'm crap at it. I nearly killed Fred this morning with a spear. And as for the gathering! I mean, if I have to drink another nettle soup, I'm going to crack up.'*

Wife: *'Brilliant, just bloody brilliant! So what are we and the kids going to do now for food?'*

Husband: *'Well, I have a plan. I was thinking of an entirely new approach to food called farming.'*

Wife: *'Never heard of it.'*

Husband: *'Well, I only just came up with the term. But I was thinking, what if we didn't have to travel for miles to catch an animal but could just walk outside and have it there waiting? What if we could somehow catch a few wild animals and become friends with them? Then we could domesticate them, keep them in an enclosed space and then get them to breed. We'd have a ready supply of animals to eat.'*

Wife: *'OK, but that sounds like an awful lot of protein. What about the berries?'*

Husband: *'Feck the berries! I have a better idea. We take a few broad-leaf grasses and cross-pollinate them so as to develop a type of cereal that you can then grow, process and eat. It'll be far more nutritious in the long term, meaning we won't have to eat nuts, and eventually we can develop it in a white powdery substance that we'll call flour, which we can use to makes cakes, have a bake-off, sell the bake-off rights across the world and then tweet about it.'*

Wife: *'Tweet?'*

Husband: *'Kind of like cave-painting but different. Anyway, farming will work out so much better in the long term and it will mean that I can stay home and help you mind the kids.'*

Of course, I am only guessing as to how the conversation went between husband and wife when human civilisation decided to leave hunter-gathering and turn its attention to a new, novel technique that would become known as 'farming'.

What we don't need to guess is what the early results of this new lifestyle looked like. For that, we just need to go to the Céide Fields, an archaeological site with a visitor centre on the north Mayo coast. Here lies not only the most extensive Neolithic site in the world but also the oldest known field systems on the planet, which dates back more than 5,000 years.

ACHiLL ISLAND

Achill is Ireland's largest island. Despite being connected to the mainland by bridge, when you're in among its vast blanket bogs, soaring sea cliffs, secluded sandy beaches, rocky headlands and rugged mountains, it can feel a world away from civilisation – despite civilisation being less than an hour away in Newport. This feeling of isolation only doubles when winter closes in.

While the island is a walkers' and cyclists' paradise, particularly during those few fine summer days, one of the most striking features on Achill is the Deserted Village at the base of the Slievemore Mountains. The Deserted Village was one of several communities, concentrated in the west of Ireland, whose inhabitants, when faced with the brute force of the Great Famine, had little option other than to pack up and leave, turning their settlements into Ireland's original ghost estates.

Today, you can visit the village and see the ruins of the 80-odd houses that still survive. And if you take part in one of the country's most scenic thirteen-milers, the Achill Half-Marathon, your route will even allow you to follow in the footsteps of those who left this village, albeit travelling a lot faster.

KNOCK

THE THREE MiRACLES OF KNOCK

1.

On 21 August 1879, at about 8 o'clock in the evening, fifteen people aged between 5 and 75 saw an apparition of Our Lady, St Joseph and St John the Evangelist in a blaze of heavenly light at the southern gable of the local parish church in the rural village of Knock, County Mayo. That is the first miracle of Knock.

2.

The second miracle of Knock is that the apparition was not attributed to the local poitín. Instead, international interest stretched from the Vatican City to Queen Victoria and after an ecclesiastic commission of inquiry was established, it was decided that the apparition was genuine.

3.

The third and final miracle of Knock is not the construction of a complex of buildings that eventually included five churches, a religious books' centre, caravan and camping facilities, museum, café and hotels to service the tens of thousands who come in pilgrimage every year (including, at least once, every Catholic child in Ireland). No, the third miracle of Knock is that, despite being located in deepest, darkest Mayo, on top of a hill in boggy, foggy terrain, an international airport was built here in 1985 that now caters for more than 600,000 passengers annually.

ROSCOMMON

At 75.6 years for men and 82.4 years for women, Roscommon's residents currently have the highest life expectancy of any county in Ireland. And before anyone jumps in to suggest that 75.6 or 82.4 Roscommon years must seem really feckin' long, if you've ever been to the place and not just through it on the way to somewhere, then you'd see that Roscommon is a great oul county to live and hang out in. And while I might not want to hang out there for a full 75.6 years myself, I always enjoy visiting.

Though it might be impossible to ever correctly identify the reason for Roscommon people's longevity, it is probably safe to say that it has less to do with their proud tradition of 'boil water' notices and more to do with their laid-back nature and how every child in the county over a certain age has spent at least one summer running the length and breadth of its beautiful Lough Key Forest Park.

BODA BORG

Situated just five kilometres away from Boyle, the funniest town in Ireland – *Moone Boy*, anyone? – Lough Key Forest Park, with its forest hikes, canopy walks, orienteering trails and bicycle loops, has been keeping Roscommon families fit for more than a generation.

Of all the attractions at the park, and there are several, my favourite has to be Boda Borg, Ireland's answer to that beloved Channel 4 game-show of the 1990s, *The Crystal Maze*. This was a game-show where teams of contestants as colourfully dressed as day release prisoners, led by a host your mother never trusted, scuttled through a labyrinth of themed zones in an attempt to collect as many golf-ball-looking crystals as they could. To get these crystals, team members had to enter small rooms and complete a task in a set amount of time. Failure to do so often resulted in the contestant being locked into the room.

Such was the popularity of the programme in the multi-channel households of '90s Ireland that, for a period of time, the most frequent call-out for fire-brigades outside of Hallowe'en was to houses where younger siblings had become trapped in bathrooms, basements, lofts and attics, locked in by their brothers and sisters after failing a homemade task. Thankfully, such incidences of familial imprisonment in Ireland have dropped since the opening of Boda Borg in Lough Key Forest Park, with more than 40 rooms of activities, challenges, tasks and puzzles for family and friends to get their *Crystal Maze* hit.

Why Roscommon is the only European location for a Boda Borg outside Sweden is unclear, but what is clear is how lucky we are Roscommon was picked. Though just as competitive and challenging as *The Crystal Maze*, the thinking behind Boda Borg is quite different: you collect stamps instead of crystals; you usually enter the games often as a whole team; there is no fear of being locked in; and there is absolutely no compulsion to turn up wearing ridiculous multi-coloured '90s jumpsuits. As a result, Boda Borg, occasionally called the 'thinking man's stag', has quickly garnered its own cult following and is one more great reason to visit north Roscommon.

OWEYNAGAT CAVE

⚡ Creatures emerging from the ground to wreak havoc on the surrounding lands

⚡ A place where people live in fear, not knowing who the next victim will be

⚡ A difficult past with an uncertain future

No, we're not talking again about cryptosporidiosis and the 'boil water' notices but Oweynagat Cave in central Roscommon, which is known as Ireland's 'Gate to Hell'. The cave is situated in Rathcroghan, not far from Tulsk (the town, not the incorrectly-spelt 12th album from Fleetwood Mac). The idea of Roscommon being the mouth of Hell might come as a bit of shock to some people (Buffy the Vampire Slayer, the cast of Stargate and Joe Public for a start), but it shouldn't, given the number of megalithic tombs dating back more than 5,000 years it contains and its history of rich Celtic myths and legends.

In Celtic mythology, warrior queen Medb of Connacht was said to have lived, ruled and watched over the area in which Oweynagat Cave sits. However, she might have overlooked this one among the many in the area, as what self-respecting warrior queen would allow a Hell Mouth open up in her backyard? While no one knows exactly how many monsters use Oweynagat Cave to commute in and out from the underworld, one mythological being that definitely resides in it is the Morrígan, a goddess of death often associated with crows, ravens, Roscommon underachieving in the football and er ... death!

ARiGNA MiNES

Looking like a set of ridiculously cool cubes dropped down from outer space, the Arigna Mining Experience Visitor Centre, at the site of the old Arigna mines, proves an interesting tourist stop in the northern-most tip of Roscommon.

In order to help visitors experience what coal mining life was like in the Arigna Valley during the mines' years of operation, between the 1700s and 1990, visitors are invited on an underground tour in some of the narrowest coal seams in the western world. Led by former miners, some of whom started working in the mines shortly after their Confirmation (at the age of twelve, for anyone who's asking), tourists are literally brought to the coal-face and shown how miners used to work, often lying on their sides as they first hacked out, then shovelled and then pulled back the coal. With lighting and sound effects to go with the dark and damp conditions, all that is missing is the fear of a cave-in and the development of a respiratory problem to complete this authentic experience.

ELPHiN WiNDMiLL

Anyone arriving into the village of Elphin would be forgiven for believing that they have just entered a little corner of Holland.

This is not due to the local inhabitants' above-average height, the village's relaxed attitude to marijuana or even the adherence of their Gaelic football team to the concept of total football (none of which are actually true, I think) but because of the beautifully restored 18th-century windmill that sits just outside the town.

Built around 1730, and wonderfully restored in 1996, this little slice of lowland Holland sits captivatingly just out from the village and is the oldest operational example of its kind in Ireland. With four timber sails and a thatched rye rotating roof, the windmill is a truly remarkable structure, especially when you consider it was constructed well before the agricultural and industrial revolutions came to England, let alone Roscommon.

SLiGO

AT ONE STAGE IN ITS HISTORY, THERE WERE ONLY FOUR REASONS YOU WENT ON HOLIDAY IN SLIGO

1.
You were too tired to continue driving to Donegal.

2.
You were too depressed to risk continuing on to Mayo.

3.
You couldn't afford Kerry.

4.
You just plain ran out of reasons not to go.

Thankfully, things have all changed and now if you are a poetry-liking, art-appreciating, big wave-surfing, beach-enjoying, mountain-enthusing, history-loving person, with a strong bladder and you don't mind listening to local radio on the drive over, then Sligo is most definitely the place for you.

It has artists of the fame of W.B. Yeats and his painter brother Jack, a history that spans from the Celtic legends Diarmuid and Gráinne to the modern-day legend, Countess Markievicz. Then there are places of natural beauty, from the waves at Mullaghmore and the beaches of Enniscrone to the spectacular weirdness of Ben Bulben, and stately grace in the form of Lissadell House. And when you add in Westlife, the dearth of public toilets and a rich history of exorcisms and burying the undead (ask Bram Stoker), then Sligo really is the place to visit.

BEN BULBEN

FAMOUS BENS

Ben Franklin: One of the famous founding fathers of the United States, who was also a renowned polymath, author, printer, politician, postmaster, scientist, inventor, civic activist, statesman, diplomat and bank-note. Crap at field sports though.

Uncle Ben: Famous rice producer, who may never have existed.

Ben Affleck: Occasionally irritating award-winning US actor/film-maker known for his success, intelligence, the odd crap movie and the fact that he appears to be a genuinely nice guy. This genuine niceness is why many people find him so irritating. Friends with Matt Damon.

Ben Stiller: Occasionally irritating award-winning US actor/film-maker known for his success, intelligence, the odd crap movie and the fact that he appears to be a genuinely nice guy. This genuine niceness is why many people find him so irritating. Not friends with Matt Damon.

Big Ben: Famous London landmark, but you can never remember if the name is for the tower, the clock, the bell or the whole thing together.

Gentle Ben: Famous bear that appeared in several American children's TV movies. Left Hollywood after becoming disillusioned at its lack of strong female characters. Currently works as a stage actor on Broadway.

Ben Hur: Popular Jewish nobleman, who, having been falsely accused of an attempted assassination and then enslaved by the Romans, manages to turn his life around, becoming a successful charioteer, witnesses the crucifixion of Jesus Christ, gains redemption and then stars in the movie *Ben Hur*, which is on TV every Easter without fail.

BEN BULBEN

The Ayers Rock of the west of Ireland. This is less due to the fact that it is held sacred to the Aboriginal people of the area than because of its striking appearance and how, like Ayers Rocks, it seems to appear out of nowhere. Best described to someone yet to visit as looking like a giant ship made of mountain that is heading for the sea.

Along with being mysteriously majestic, it is also the only known Irish home to the fringed sandwort, a perennial herb, which appears to have hitched a ride on the mountain some 100,000 years ago on its way to Greenland and then forgotten to get off.

THE MERMAID STONES

The story goes that Thady Rua O'Dowd, chief of his clan, was looking for love and, after several failed first dates, was about to give up when he rounded a bend in Scurmore, south of Enniscrone, to find a mermaid combing her hair as she sang. Mermaids were a lot more common back then, before the introduction of drift nets, although coming across one so far ashore was rare.

While it was love at first sight for dear Thady, he was not so sure how she would react to him. So he decided to sneak up on her and rob the fish-tail cape she sat on, a magical item that allowed her to switch from sea to land and back again. This diet-kidnapping worked and once Thady had her cape, Eve, as she became known, agreed to marry him!

For several years they lived happily, helped by the seven merry children that they had together and possibly a small bit of Stockholm syndrome on Eve's part. All the while, Thady kept Eve's cape hidden until, one day, when he was in the process of moving it to a safer spot ahead of a business trip away, his youngest child saw him. After the child innocently told his mother its location, Eve felt compelled to retrieve the cape and when she had, felt compelled to skedaddle back to the sea. However, as she approached its shore, she realised she couldn't bring all of her children with her. So taking just one and not wishing the others to follow her into the sea and drown, she turned them into six granite stones instead, an act that probably made the Children of Lir feel they got away lightly!

To this day, the stones remain, though you'll need a bit of luck to find them. Five of them lie together in dense undergrowth and trees along the coast road three kilometres south of Enniscrone. The sixth stone sits a little closer to the water, keeping watch over Killala Bay. All six are hoping that one day their mother will return and release them from their granity state so they can head off to college or something, because it's feck-all craic being a stone.

MULLAGHMORE WAVES AND STRANDHILL

There was a time when surfing in Ireland was as alien to the country as a month-long dry spell, space travel and well ... aliens.

However, that changed with the advent of really good wet-suits and the trail-blazing antics of a small and select group of vision-aries. These few looked out to the ocean and noticed the currents, not the cold; the swell, not the squalls; the opportun-ities, not the awful weather. And pretty soon, surf-schools started to emerge up and down the Wild Atlantic Way from Easkey to Enniscrone, Lahinch to Rossnowlagh, and Bundoran to Barleycove as hardy men and women wrapped up in wet-suits and took to the water to try and catch that wave. And that's how we got a surfing scene and became known as 'Europe's cold-water Indonesia'.

Two Sligo destinations that epitomise this western surf culture are the village of Strandhill along Sligo's southern coastline and the waves off the headland of Mullaghmore further north.

The first is one of the original meccas of Irish surfing. While it will never be as exotic as Maui, as ripped as Byron Bay or as sun-kissed as Malibu Beach, it boasts a mid-summer's sun that stretches the day almost to dawn, so for those who surf, those who like to watch and those who just have curly blonde hair and wear shell necklaces, Strandhill is the place to be.

The second place is Mullaghmore, which is perhaps Ireland's best known big-surf hide-out. Here, when the winter weather really tears it up and every sane person and their dog are taking shelter beside a fire and a pint, surfers who have trained for years do battle with these great salty waves, which rise up to fifteen metres high. And when they catch the barrel of one of these behemoths, those watching from the Mullaghmore headland can only look on in awe and wonder, that if they are going to surf in this cold weather, the least they could do is throw on a coat.

THE GLEN

The Glen is one of the most magical hideaways in Ireland and one of Sligo's best-kept secrets.

In the heart of the Coolera Peninsula, running for over a kilometre, 20 metres deep and twelve metres wide, the Glen is one majestically long cleft in Knocknarea Mountain, said to be formed by some ancient eruption or earthquake that parted the side of the mountain, tore apart veins of limestone and filled the gaps full of fairies.

Filled with every sort of vegetation, the Glen could be the backdrop to an Indiana Jones movie – it would be no surprise if the only thing that disturbs you, as you wander among its holly and honeysuckle, bramble bushes and beech, was the soft rumble of a giant boulder rolling down towards you as you escape with a golden idol.

Of course, you are unlikely to find anything like a golden idol here but then for a long time you were unlikely to find anything here at all, so secret was the Glen to anyone outside Sligo. In fact, such was the omerta of silence that surrounded the Glen that any local found telling a non-Sligo person about it risked inviting a fatwa against them or, worse still, being exiled off into Leitrim.

While this has changed and tourists are now welcome to come and visit the Glen, finding it can still prove a challenge with the best directions a Sligoman gave me being *'head out towards Knocknarea until you get onto a steep slope overlooking the sea. Park here and then continue along the road until you get to the second tree on the left. You can then climb across the fence here, head through a gate and you'll find it'*. After that, I was on my own.

~

MUNSTER

~

CLARE

THE FOLLOWING DIALOGUE SHOULD BE SPOKEN LIKE JOHN CLEESE AKA REG FROM THE MONTY PYTHON CLASSIC *THE LIFE OF BRIAN*

Reg: '*What has Clare ever done for Ireland?*'

Crowd: (Silence)

Crowd member 1: '*The Cliffs of Moher.*'

Reg: '*What?*'

Crowd member 1: '*The Cliffs of Moher.*'

Reg: '*Oh yeah, yeah, they did give us that, that's true.*'

Crowd member 2: '*And the Burren.*'

Reg's colleague: '*Oh yeah, the Burren, Reg, remember what the countryside used to be like.*'

Reg: *'Ok, right, alright, I grant you the Burren and the Cliffs of Moher but aside from those two things, what has Clare ever done for Ireland?'*

Crowd member 3: *'The Aillwee and Doolin Caves.'*

Reg: *'Obviously the caves, I mean the caves go without saying, don't they? But apart from the Cliffs of Moher, the Burren and the caves ...'*

Crowd member 4: *'Poulnabrone Dolmen.'*

Crowd member 5: *'Bunratty Castle.'*

Crowd member 6: *'The beaches of Lahinch.'*

Reg: *'Yeah, alright, fair enough.'*

Crowd member 7: *'And the likes of Brian Boru, Michael Cusack, Edna O'Brien and Keith Wood.'*

Reg's other colleague: *'Yeah, Reg, they're people we'd really have missed if Clare had left.'*

Crowd member 8: *'Shannon Airport.'*

Crowd member 9: *'Ardnacrusha Dam.'*

Crowd member 10: '*The festivals of Lisdoonvarna, Miltown Malbay and Ennis.*'

Reg's colleague: '*It is great craic in the summer.*'

Reg's other colleague: '*They certainly know how to have fun.*'

Reg: '*Alright, but apart from the Cliffs of Moher, the Burren, the caves, the dolmen, the castle, the beaches, the noteworthy people, the dam, the airports and the festivals, what has Clare ever done for us?!*'

Crowd: (Silence)

Crowd member 11: '*A Clareman invented the submarine?*'

Reg: '*Oh submarine! Shut up!!!*'

THE BURREN

With *'not water enough to drown a man, wood enough to hang one, nor earth enough to bury them'*, the Burren was never going to make the top 100 places to hide a genocide. While Cromwellian soldier Edmund Ludlow may have meant the above remark as a criticism, what he failed to realise is that the Burren, far from being a hell of sorts, is an incredibly beautiful topographical region unlike any other.

Most accurately described as otherworldly, with its lime-stone pavement criss-crossed with large cracks known as grikes, what makes the Burren truly unique is that three-quarters of the species of flower in Ireland only grow here in this 250-square-kilometre region of stark karst land-scape. And what is more, due to its unusual environment, the region supports Arctic, Mediterranean and Alpine plants side-by-side in perfect harmony. I bet Ludlow didn't notice that either!

THE AiLLWEE CAVES

SEVEN MODERATELY iNTERESTiNG FACTS ABOUT THE AiLLWEE CAVES

1. While only 300 metres of the Aillwee Caves are open to the public, the actual Aillwee Cave system is over a kilometre in length.

2. Most people spell Aillwee Caves with just one 'l'.

3. The remains of brown bears have been found in the caves, leading experts to believe that the caves served as a den for them. Some have gone even further, guessing that the Aillwee Caves were one of the very last places that bears lived in Ireland, making it an Alamo of sorts, where the bears made their last fateful stand against early Irish settlers.

4. Father Ted once got lost in the Aillwee Caves.

5. Father Ted couldn't believe that the caves were 15 million years old and he was probably right. In fact, parts of them are up to 350,000 years old. To put that in context, human beings only began making fire 350,000 years ago. And to put that in context, that is some 349,995 years before we began to Instagramming ourselves making a fire to impress our friends!

6. The caves were rediscovered in modern times by a local farmer back in 1944, when he followed his dog down what at first he thought to be a rabbit-hole. He didn't tell anyone about the caves for nearly 30 years, leading some to suggest that it took him that length of time to get over the shock of also finding the Mad Hatter, March Hare and Queen of Hearts while he was down here.

7. One amazing feature of the caves is its huge stalactites, which stretch down from the roof, and the stalagmites that stretch upwards to meet them. And anyone who has ever set foot into an Irish geography classroom will know how to tell the difference between these, because 'tights fall down'.

CLIFFS OF MOHER

Almost everyone in Ireland knows how beautiful and stunning the Cliffs of Moher are. Rising up to 214 metres straight out of the wild Atlantic Ocean, the views from the cliffs of the Aran Islands, the Maumturks and the Twelve Pins mountain ranges to the north, Loop Head to the south and the vast Atlantic Ocean are simply breath-taking.

While it should come as no surprise that the cliffs regularly feature on top of the list of the most popular tourist destinations in the country, lesser-known accolades that they have picked up include:

★ Ireland's Loudest Place to Propose to a Loved One

★ The Nation's Most Expensive 1-Hour Car-Parking

★ The Country's Worst Tourist Site to Visit Straight after Getting Your Hair Styled

★ The Country's Best Tourist Site to Get a Blow Dry

★ Ireland's Most Absurdist Use of Double-Yellow Lines

★ The World's Least Observed Safety Notice

ARDNACRUSHA

In the last 30 years of Irish politics, you could probably count using just your fingers the number of times a government took a policy decision for the greater good that you could really call 'gutsy'.

Back in the day, it was a little different as government ministers made big decisions for our nascent country every other day. Perhaps one of the gutsiest of these was Ardnacrusha or the Shannon Scheme when, just three years after gaining independence, the Irish government decided to spend more than a fifth of its entire annual budget to build the world's largest hydroelectric station!

Elbowing its way onto the government's agenda between 'The continuing threat of anti-Treaty forces' and 'Is it too early to enter Eurovision?', Minister for Industry and Commerce Patrick McGilligan's Shannon Scheme must have raised eyebrows. One might only wonder at the conversation.

Patrick McGilligan (PMcG): *'So gentlemen, I won't keep you long. Basically, in order to electrify the country, I propose* *building the largest hydroelectric station on the planet. It will be based on the Shannon and cost over 20% of our national budget. Any questions?'*

Minister 1: *'Will it affect Clare's chances in the hurling?'*

PMcG: *'No, we'll get Siemens and the Germans to build it so it won't affect training.'*

Minister 2: *'Anything about this that is against Catholic teaching?'*

PMcG: *'Nothing against hydroelectric stations in the Bible, as far as I know.'*

Minister 3: *'How much will it cost again?'*

PMcG: (Doing some quick calculations on the back of an envelope. Someone murmurs *'carry the 2'*.) *'From our estimates, a fifth of our annual state budget.'*

Minister 2: *'Jaysus!'*

Minister 3: *'Jaysus!'*

Minister 4: *'Jaysus!'*

Minister 5: *'Jaysus!'*

Minister 6: *'Feck off with ya!'*

Minister 1: *'Will that affect Clare's chances in the hurling?'*

PMcG: *'No more than any other county. C'mon lads, it'll be grand. All in favour?'*

Ministers 1–6: *'Aye.' 'Alright then.' 'No bother.' 'Grand.' 'Fine so.' 'C'mon Clare boy!'*

W.T. Cosgrave: *'Right so, world's largest hydroelectric scheme is passed. Next item on the agenda: the Eurovision.'*

CORK

If Ireland were ever invaded, Cork would likely be the last place to fall. Instead of *Battle: Los Angeles*, it would be *Battle: Youghal* and you can be sure that local resistance would be tough. Cork is the original rebel county, holding out against the shampoo-loving Vikings when everywhere else in the country fell.

On the flipside, if ever a county were to secede from the Republic and declare independence, it would also most likely be Cork – feasibility studies on how it would sustain itself financially are rumoured to have already been carried out. But that's Cork for you and that's why we love them. Or part love, part hate, anyway. And the feeling is mutual.

The thing about Cork is that they often consider them-selves quite different from everywhere else. As one of their patron saints, Roy Keane, once said, '*Corkman first, Irishman second*'. And to be fair, you can sometimes under-stand why. I mean, Cork city is about as far away from Dublin as you can get. And the wild, rugged and really quite wonderful west Cork is nearly as far away from Cork city as you can get. So that's pretty far away! And while it might not be the most populous county in Ireland, it sure is the biggest, so no wonder its people sometimes view the place as less a county and more a sovereign region in its own right: the People's Republic of Cork. But we let them go on this; after all, this is the first place in Ireland the potato was planted. That has to be worth something.

THE ENGLISH MARKET

Though the English Market in Cork city centre might not be as exotically titled as its foreign cousins such as the Chandni Chowk in Delhi and the Grand Bazaar in Istanbul, it is every bit as appealing in that hell-of-a-lot-smaller-but-still-sensuously-enticing sort of way. This is due to the wide range of champion artisan food producers who have joined the long-held family-run food-stalls to create a culinary utopia, with foodstuffs that contain the freshest of recently landed fish to the choicest locally reared meat and from the crunchiest baked breads to the smelliest of cheeses (which, I am informed, is a good thing).

It wasn't always this way, however, and many in Cork might remember a time when the market was renowned for its short-cuts, cheap-cuts and NAFF jacket off-cuts that used to characterise it up until the 1990s. Since then, it has grown as a place of great culinary esteem, joining the town of Kinsale and Ballymaloe Cookery School as evidence of Cork's status as a gastronomic mecca. And if you won't take my word on this, you should take your cue from Queen Elizabeth II, who famously dropped into the English Market during her visit to Ireland in 2011, and was rumoured to have picked up some haddock for supper.

FOTA WiLDLiFE PARK

Many of the birds, reptiles, mammals and marsupials that live in Fota Wildlife Park are separated from the general public by a mixture of wooden enclosures, water trenches and the type of electric fences your uncle keeps the cattle in with. However, if the animals really wanted to escape, it wouldn't take a helicopter, some bolt cutters and a box of dynamite for them to get far. I mean, there's even a train service that runs alongside the place that they could hop onto! But they don't; instead, they seem perfectly at ease in this relaxed, stress-free environment on the outskirts of Cork city.

Since opening in 1983, Fota Wildlife Park has become home to nearly 30 mammal and 50 bird species. Not only do they include your usual suspects like giraffes, ostriches, kangaroos and zebras, they are also home to more exotic-sounding creatures, from grey-cheeked mangabeys and white-faced sakis to siamang gibbons and red-ruffed lemurs – and let's not forget those mischievous-looking meerkats and everyone's favourite two-foot-tall rodent, the capybara!

While there are animals like cheetahs, Sumatran lions and the odd Indian rhino that are a little bit more enclosed from the visitor than most, what typifies the park is how many of the animals roam freely among the strolling visitors – so much so that it is not unheard of to find yourself queuing behind an Indian peafowl at the Savannah Café or being served by a Bennett's wallaby in the Serengeti Gift Store.

BLARNEY CASTLE

Every year, Blarney Castle makes it into the top ten most-visited tourist destinations in Ireland, welcoming over 300,000 visitors. Most of those who come do so to be held by a burly man as they bend over backwards and kiss the famous Blarney Stone. In doing so, they hope to gain the gift of the gab, which is meant to allow them to speak with eloquence, fluency and the minimum amount of errrs, emmms and hmmms.

While most people know what you'll acquire by kissing the Blarney Stone, its origins are less well known. There are several legends:

> Some think it was a pillow used by Jacob, the father of the Israelites, who preferred a head rest made of limestone over feathers. Brought to Ireland by the prophet Jeremiah, it became known as a *Lia Fáil* or 'Fatal Stone' and was used as an oracular throne by Irish kings.

> Others believe it was the pillow that St Columba died on on the island of Iona. Any wonder, with the chill it must have given off. On his death, the stone was sup-

posedly removed from Iona to mainland Scotland and became known as the Stone of Destiny, until Robert the Bruce sent a portion of the stone to Ireland to thank the Irish for their support in his emphatic victory over the English at Bannockburn. So thanks, Rob, for that one – although we'd probably have preferred the cash.

❧ There are those who argue it is in fact the Stone of Ezel, brought back from the Crusades. This stone is meant to be what the biblical character of David hid behind as he attempted to flee his enemy Saul. That said, if you've seen the size of the stone, you'd have to wonder whether David was really feckin' small or just terrible at hide-and-seek.

❧ Then there are rumours it was stolen from Stonehenge and brought to Ireland on the medieval version of the Stena Line ferry.

❧ And finally, and perhaps most controversially, there are a few who believe it is just a big feck-off rock of carboniferous limestone that put at least one worker's back out while the castle was being built.

CASTLEFREKE

CASTLE-WHAT? CASTLE-WHO? CASTLE-WHERE?

Indeed, all very relevant questions, and questions that anyone who ever played that classic Irish educational board-game *Discovering Ireland* as a ten-year-old should be able to answer. Castlefreke is, along with Dublin, Belfast and Cork, one of the 50 most important towns and cities in Ireland – at least if you played *Discovering Ireland*, you were brought up to think so.

The aim of the board-game was to successfully negotiate road-blocks and missed turns to visit several places in Ireland in the shortest possible time. To add education to the fun, each of the towns and cities that you could end up in with had a wonderful little description attached, which you could read as your sister sent you to Sligo or road-blocked you on the way down to Newcastle West.

For instance, Castlefreke's fellow Cork location Bantry was described as *'positioned beneath sheltering hills at the head of the famous Bantry Bay, one of the most beautiful bays along the Irish coast'*, while up the road in Youghal, *'famous for its point lace'*, we learnt how Sir Walter Raleigh, once mayor of the town smoked the first tobacco and grew the first potato here.

Into this mix came Castlefreke, which every ten-year-old in the country, except possibly those living in Cork and probably those living in Castlefreke, thought was a place of extraordinary note and very much worth visiting.

In hindsight, there were two clues that, although charming, picturesque and probably a nice place to stop for tea, Castlefreke was a lot less important than its board position suggested. The first was the fact that Castlefreke happened to be where the board-game was manufactured. The second was that its description included statements such as *'chequered history, much of it violent'*, a castle *'destroyed by fire in 1910'*, *'rebuilt with insurance money'* and finally *'bought during the depression of the 1950s, its lead, timber and artefacts dismantled and sold'*!

(Unsurprisingly, Castlefreke didn't make later editions.)

KERRY

When it comes to the wilds of nature, most counties have something of value but in terms of sheer volume, Kerry stands out. With more places of outstanding natural beauty than it knows what to do with, it is little wonder that when people visit Ireland, many come just for Kerry.

With all this natural beauty, it is no surprise that, while Kerry's southern neighbour Cork might sometimes think of itself as a separate country to the rest of Ireland, Kerry kind of is. They speak a little differently, they act a little differently and they play football a little differently, which their record tally of All-Ireland titles attests to. Add the fact that they are at least 20 minutes closer to the sunset than the rest of us, it is small wonder that the littlest bit of resentment is harboured by those of us who do not hail from the Kingdom.

However, any jealousy will quickly vanish if you pay a visit, for, pulling in alongside the road to watch a summer sunset across the Dingle Peninsula, you'll quickly realise that you can't possibly have room in your heart for envy when confronted by beauty like this.

KiLLARNEY NATiONAL PARK

Located beside the town of the same name, Killarney National Park is a great place to start your visit to Kerry. Set over 100 square kilometres, the park is chock-a-block with stunning sites, such as Muckross House, Ross Castle, the Purple Mountains, Torc Waterfall and the lakes of Killarney to name just a few. While all these captivate, perhaps the most interesting part of the park is its sheer spread of flora and fauna, making it the Amazon of Ireland. With anywhere between 85% and 112% of our native species to be found here, it is a naturalist's dream.

For tree lovers, there are oak, yew and wet woodlands that make up the most extensive area of natural cover in the country. Above 200 metres, its mountains turn to blanket bogs and wet heaths, with enough mosses, heathers, lichens and liverworts to keep bog followers up all night. Animal enthusiasts, bird buffs and fish fans will be stirred by every shade of Irish wildlife from red deer, red squirrel and white-tailed eagles to wood warblers, Arctic char and

Killarney shad, as well as the king of them all, the Kerry slug, an invertebrate that is reputably the only slug on the planet that can roll itself into a ball! Good times!

In fact, the only thing that takes away from the wonder of Killarney National Park are the bright pink rhododendron bushes that seem to sprout up everywhere, like the most annoying house-party guest ever! While the rhododendron might at first look kind of cool, they are the type of guest who takes over the music selection, eats all the sausage rolls, spills their wine on the carpet, throws up in the sink and then shifts your sister. First introduced back in the 19th century by some bright spark who thought they'd bring a bit of craic and colour to the grounds, 150 years later and despite every possible hint from telling them to feck off home to taking a lump out of them with the chainsaw, Killarney still can't get rid of them. So if you do see any when you are down there, best steer clear and whatever you do, don't invite them back to your place.

SCEILIG MHICHÍL

Just off the coast of Kerry's Iveragh Peninsula lies the island of Sceilig Mhichíl. It is one of two prominent land masses that emerge from the ocean and that are popularly known as the Skelligs. What sets it apart from its smaller nearby sibling, Sceilig Bheag, are the Christian monastic settlements that lie near its summit. Residing at the top of a steep stone stair, these structures are characterised by their beehive design and were built early in its 600 years of existence as a monastery between the 6th and 12th centuries.

On the bouncy boat ride over from Portmagee, you might wonder how a group of monks, led by St Fionán, said to be the monastery's founder, decided whether they would move to Sceilig Mhichíl, but it was surely by way of a list of pros and cons.

PROS:

---- Peace

---- Solitude

---- One of the world's largest gannet colonies next door on Sceilig Bheag

---- Great drying for the clothes

---- Peace

---- Hundreds of puffins, those colourful crazy-looking beaked birds, for company

---- Low smog levels

---- Great views of the whole world during those few annual days of Atlantic sunshine

---- The perfect hideaway for a Jedi Knight trying to keep a low profile

---- Peace

CONS:

---- Nearly twelve kilometres off the mainland in the middle of the Atlantic Ocean

---- Atlantic storms

---- No shops nearby

---- Limited parking

---- No chance of rescue

---- Atlantic storms

---- Terrible mobile phone coverage

---- Bird poo

---- Atlantic storms

Whether it was the peace or the puffins that swung it in the end, they decided to make the move. And if you don't mind heights, sea travel and the wet ankles you might have to endure to get out here, reaching their simple but stunning beehive huts on one of those rare sunny days, you'll realise it was the right decision.

DiNGLE

In 2005, controversy hit west Kerry's most popular tourist town, Dingle, when the then-Minister for Community, Rural and Gaeltacht Affairs decided that, owing to the fact that the town of Dingle was located in the Kerry Gaeltacht, it was only right to change its name to *An Daingean*.

Confusion soon followed, with tour operators finding it hard to locate. Then tales started to circulate about bus-loads of American and German tourists ending up in the north-eastern Offaly town of the same name. Finally, when it was rumoured that the town's celebrity dolphin, Fungi, had been found, lost and confused, a few kilometres outside Tullamore, having become mixed up on his way to work, the residents of Dingle finally had enough.

So in October 2006, the town held a vote and 90% of them voted for the town name of Dingle to be reinstated and the future of tourism in this little gem was guaranteed forever. Hooray!

Of course, the village is much more than a name change. It is a wonderfully compact Kerry town characterised by its hilly streets and the fishing port that opens out onto the Atlantic. It is a plethora of colourful cafés, pubs, shops and stalls that offer a wealth of choice, character and charm. And it's a town full of vibrant festivals and animated traditions.

The old traditions include the Wren Day celebrations when the most motley of groups of costumed creatures imaginable sing, dance, drink and drum their way through St Stephen's Day, turning Dingle into a damp, cold and chaotic version of the Rio carnival, while its festival range from food to film and include Other Voices, when several leading lights in the Irish and international music scene gather in the intimate setting of 200-year-old Church of St James for three nights of music.

Finally, Dingle is also about its uniqueness, about how you can walk into a pub and be met by two old men and their dog out front and a Caribbean night in full flow out the back. Why? Because that's Dingle for you.

THE RiNG OF KERRY

FAMOUS RiNGS

Ring: Officially known as An Rinn, this is the Waterford Gaeltacht. You didn't know there was a Gaeltacht in Waterford, did ya?

Christy Ring: If not Ireland's best ever hurler, certainly Ireland's best ever hurler that no one under the age of 60 has ever seen.

Engagement ring: Precursor to marriage and something that should never, ever, ever be picked out by the groom-to-be.

The Lord of the Rings: Classic trilogy of books written by J.R.R. Tolkien sometimes seen as a forerunner to *Harry Potter* except with more hobbits and fewer crap games that don't make any competitive sense (we're looking at you, Quidditch)!

The Ring: Japanese horror movie that should never be left out accidentally at a slumber party for eleven-year-olds.

The Ring of Fire: What Johnny Cash fell into during his 1963 song of the same name. The Ring of Fire either refers to falling in love or a female body part best left unexplained. It certainly does not refer to an outdoor barbecue accident.

The rings of Uranus: Both an astronomical feature around the planet Uranus and the reason why more than a dozen teenage boys get detention every year when they put this down as an answer in biology tests.

THE RING OF KERRY

Pound for pound perhaps the most scenic route in Ireland. Tracing the coastline of the Iveragh Peninsula and passing picturesque towns and villages of the likes of Sneem, Glenbeigh and Kenmare, the Ring of Kerry is one jaw-dropping vista after another. At 179 kilometres long, Ireland's most popular road-trip usually takes a leisurely four hours by car, five if you get stuck behind a tourist bus, six if you are on that bus or just over two if you take some of its 100 km/h speed-limit signs seriously and manage not to end up the far side of a ditch. One word of warning: unless you want to see the whites of an Irish bus-driver's eyes as he veers his coach towards you, busy pointing out a shrine to a group of Nebraskans, I'd recommend you drive the Ring of Kerry counter-clockwise.

LIMERICK

Limerick is a little like the Keyser Soze of Irish counties, with a name far more notorious than the place ever was. Due to this notoriety, Limerick is perhaps the most misunderstood place in the country.

To start with, the city is sometimes seen as a lawless, gang-ridden, poor-at-poetry type of place – but that is in fact Waterford, where you have a much better chance of being shot or beaten, according to current statistics.

Due to it being in the west and because its Irish name, *Luimneach*, means 'bare land', Limerick is sometimes thought to be a somewhat barren place. However, once again, you'd be mistaken as the county is in fact part home to the Golden Vale, Ireland's most fertile land.

With it being in the west of Ireland, you might think it is a wet place – but it was in fact the setting for Ireland's longest drought when, between April and May 1938, there were 37 days without a drop of rain!

And although the county might today be considered a bastion of middle-politics, most people aren't aware that Limerick was, for a while, Ireland's only ever Soviet republic, albeit for twelve days in 1919!

The thing is, Limerick surprises. With an illustrious history of playing a significant role in most of Ireland's wars, revolutions, rebellions, sieges and uprisings, beautiful rolling landscapes, a wealth of culture (it was Ireland's first national City of Culture in 2014) and locals who are almost as friendly as they are mad for sport, Limerick is sure to amaze any visitor.

ADARE

Based on entirely spurious statistics, it is believed that between 40 and 50 weddings receptions a week take place in the village of Adare, with that figure jumping to between two and three an hour at the height of the summer. Indeed, there is almost certainly someone getting married in Adare right now!

It is no surprise that Adare is perhaps the most popular place in the country to wed. First of all, it is the ideal geographical compromise for the couple who are half-Dub and half-not-Dub. Second, the village has to be one of the most appealing little settlements either side of the Shannon, with half of the parish covered with either thatch or Virginia creeper. Then you have any number of old ruins that wedding photographs can use as back-drops, from Desmond Castle and the Franciscan Friary to Adare Manor and the Trinitarian Priory that now plays home to the local Catholic church. And lastly, if ever there was a place to quell a marital argument with a memory – '*but sure wasn't our wedding wonderful?*' – Adare is that spot.

A final surprising fact about the village is that, despite its beauty, charming personality and great sense of humour, Adare itself is still single!

THOMOND PARK

Situated in the city of Limerick, west of the Shannon and with a capacity of 25,630, the stadium of Thomond Park has long been an institution for Munster rugby. From the heady days of 1978 when between 5,000 and 50,000 people came to see the province beat the great All-Blacks 12-0 to more recent feats of marvel against the like of Gloucester and Northampton in the Heineken Cup, Thomond Park has few rivals when it comes to cauldrons of noise on match-day weekends.

As well as offering fascinating spectacles of rugby, Thomond Park inspires the five basic pillars of Munster rugby, which every Munster fan should perform.

1. PILGRIMAGE: At least once in your lifetime, taking the trip to Thomond Park.

2. CHARITY: Throwing a few euros into a bucket for a local hospice outside the stadium or buying a couple of lottery tickets from the local U-15s in the pub before the match.

3. FASTING: Waiting until the half-time whistle and the Munster penalty kick that will bring the sides back to level before you rush out for a pint and a hot-dog.

4. PRAYER: Rosaries and prayers to Paulie, Rog, Strings, Zeebs or whichever Munster saint past or present you see fit when you are down by two converted tries with just ten minutes to go.

5. FAITH: Believing that Munster will pick and drive, pick and drive, pick and drive to score that fourth bonus-point try with the clock firmly in red.

THE TREATY STONE

In fairness, you can see where Patrick Sarsfield was coming from at the time. It was 1691 and the Williamite War he had been fighting in had gone on for nearly two years. He had held the Williamite forces at bay during the First Siege of Limerick with a little help from the women of the city, who threw stones and bottles at the attackers who managed to breach the walls, but he was once again facing a siege from a larger army with no end in sight.

His leader James II, or James the Shite as he was now affectionately known, had long since fecked off back to France and the Jacobite army he left behind to fight alongside Sarsfield had already been defeated in several battles from the Battle of the Boyne to the Battle of Aughrim. If that wasn't bad enough, the French had just closed the doors to Limerick on 800 retreating Irishmen, locking them outside the city during the Second Siege of Limerick and ensuring their doom!

So Sarsfield did what he needed to do. He gave the French leaders still in command of Limerick their P45s and started negotiations with the Williamites.

It got off to a good start when he received exceptionally favourable peace terms guaranteeing full legal religious and land rights to Catholics as long as they swore an oath of loyalty to William. Result! The problem was not only did he not have a scrap a paper to write this down on, his pen had run out. So he did the only thing that came to mind and agreed to hew the treaty terms onto a large piece of limestone rock at the Clare end of Thomond Bridge.

In hindsight, you can see where it all went wrong. I mean, did he really believe that the Williamites would stay true to terms that had been scraped onto a stone previously used for mounting horses?

Of course, the terms of the agreement would probably not have been kept even had they been written in William's own blood on the finest Egyptian papyrus. And as a result, Catholic discrimination in Ireland continued for another two centuries. However, at least the Treaty Stone survived intact. And to this day, one of the more important, though slightly uninspiring-looking, pieces of Irish history still rests on a plinth on Thomond Bridge, a constant reminder of foreign betrayal, the importance of a pen and paper, even during war, and the proud fighting that the people of Limerick did to defend their city.

THE GRANGE STONE CIRCLE

MEGALITHIC MONUMENTS OF LIMERICK THAT NEVER MADE IT OFF THE DRAWING BOARD

≥ The Rathkeale Stone Rhombus ≤

≥ The Cappamore Stone Parallelogram ≤

≥ The Newcastle West Stone Trapezium ≤

≥ The Croom Stone Oblong ≤

≥ The Castletroy Stone Concave Hexagon ≤

≥ The Dooradoyle Stone Irregular Quadrilateral ≤

≥ The Bruff Stone Dodecahedron ≤

ONE MEGALiTHiC MONUMENT OF LiMERiCK THAT DiD MAKE iT OFF THE DRAWiNG BOARD

⋛ The Grange Stone Circle ⋚

Built in and around 2200 BC, the Grange Stone Circle in the east of Limerick county is one of Ireland's most impressive examples of what became the geometric shape of choice for megalithic monuments.

For a time in the early Bronze Age, such was the popularity of stone circles in Ireland that no community could do without one, in much the same way that any truly Irish village today needs a pub, a church and a GAA lotto draw. And while this led to plethora of stone circles across the island, with a diameter of 45 metres, the Grange Stone Circle is Ireland's largest examples of this type of megalithic site and as good a place as any to see one of these impressive ancient monuments.

TIPPERARY

It's a long, long way to Tipperary.
It's a long way to go.
It's a long way to Tipperary
To the sweetest girl I know!

Though this little ditty rose to prominence during World War I, it has witnessed a recent renewal in relevance, particularly for those who are now living the far side of Melbourne and whose 'sweetest girl' has moved on and in with a computer technician from Toomevara.

This 'sweetest girl' aside, Tipperary is also home to a surprising number of very important things, including:

- The birthplace of the GAA, founded in Hayes Hotel, Thurles in 1884

- The Garda Síochána College in Templemore, the training headquarters for those great men and women in blue who, despite pronouncing 'vehicle' differently to the rest of us, look out for the country all year round

- The starting point of Irish independence, with the first shots of the War of Independence occurring in Soloheadbeg in 1919

- One of the original Irish ciders, Bulmers, made in Clonmel

- The world's largest private thoroughbred breeding operation in Coolmore Stud, the Playboy Mansion for horses.

THE ROCK OF CASHEL

It's next to impossible to take a picture in or around the town of Cashel without the Rock photo-bombing into the background (the Rock in question being the 60-metre high mound of limestone covered in monastic buildings and not ex-pro wrestler and film actor Dwayne Johnson).

And while there are plenty of other alluring things to Cashel, such as its fortified townhouses, beautiful Georgian cathedral, spacious central plaza, intriguing Bolton library and charming Victorian town centre, it is the Rock that is the real draw to this south Tipperary town.

The history of the Rock is quite interesting, having apparently started life as part of a cave in which Satan was squatting at the nearby Devil's Bit Mountain. Legend has it that the Rock ended up being cast out by accident at the same time St Patrick was evicting the Dark One for rent arrears. Landing at its present spot in Cashel, it soon became the traditional seat of the kings of Munster before changing hands in 1101 when the Church took over.

Their tenure saw the construction of so many wonderful buildings on top of it that it soon became one of the finest examples of medieval architecture in Europe. And while there may have been some medieval architectural traditionalists who thought building a cathedral beside a chapel beside a 28-metre-tall round tower, not to mention the castle, might have been a bit over the top, several hundred years later it seems to have worked. For when the Rock of Cashel is not half-covered with scaffolding, it is one of the most stunning tourist sites in the country, perched high above the town where it commands spectacular views over the surrounding Tipperary countryside.

GARDA SÍOCHÁNA COLLEGE, TEMPLEMORE

In 1964, the Gardaí Training College was relocated from the Phoenix Park to McCan Barracks, Templemore, quickly establishing the town as one of the safest places in Ireland.

Since then, the college has excelled in turning out the very finest of men and women in blue, whose job it is to serve and protect our communities. While there have been many seminal moments in the proud history of Templemore College, some in its more recent have included:

2001: Height restrictions to joining the force are finally abolished. Despite initial fears, new applicants are not over-whelmingly Ewoks, Borrowers or hobbits.

2003: Irish classes are extended to include the phrases '*Tá Gaeilge agam ón cliabháin comh maith*' ('I speak Irish fluently too') and '*anois, ar mhaith leat trí pointe pionós* *don solas bríste, a pleidhce?*' ('now, how would you like three more penalty points for that faulty tail-light you cheeky git?') to meet the rise in the number of Gaeilgeoirs trying to avoid traffic violations by speaking solely in Irish.

2006: The Garda Síochána Reserve is created and also begins training in Templemore. Kind of like the Garda B-team in a game of backs and forwards, not only do they perform similar roles to the Gardaí proper such as manning checkpoints, staffing duty and preserving crime scenes, they even get their own Garda uniform and can also get into Coppers for free.

2007: The extendable baton is introduced. Early worries that it will take Gardaí several weeks to get up to speed with its usage and several months before they stop using the term 'extendable baton' as some sort of innuendo when conversing with members of the general public prove groundless.

2012: Evening classes in Twitter etiquette and 'How to make traffic updates more interesting by linking them to popular culture' are introduced as An Garda Síochána takes to the Twitter machine and becomes an immediate hit.

Though the college is technically not open to the public unless you are attending a graduation or are playing them in football, if you come in holding a bag of sandwiches from the nearby Centra or alternatively a toasted sandwich from Polly's Pub, you should get a good half-hour wandering around the old barracks, before you get asked to move along now.

MITCHELSTOWN CAVES

Discovered – it would seem, like most caves – by a farmer who was out looking for something else, Mitchelstown Caves easily fall into the category of 'the most beautiful limestone caves in Europe'. Indeed, it is no wonder as they extend sixteen kilometres one way and five kilometres in the other, with a vast array of spectacular stalactite and stalagmite geological features including the ten-metre Tower of Babel. As well as this, they do tick another more unusual box, the 'most impressive cave venue for a musical concert' category.

Though you wouldn't think putting on a public concert 60 metres below ground would be a recipe for success, it actually works exceptionally well. And for the past several summers, Mitchelstown Caves (very much in Tipperary despite their Cork-sounding name) have hosted several artists in concerts that have all been high on intimacy and atmosphere and low on portaloos and puke.

SEMPLE STADIUM AND HAYES HOTEL

It can sometimes pain Tipperary hurling fans to see their eastern neighbours Kilkenny suck so much air out of the room when it comes to their shared love of hurling. However, Tipperary can always draw comfort from the knowledge that their county plays home to two of the most important icons of the GAA, Hayes Hotel and Semple Stadium both situated in Thurles and barely a kilometre from each other.

HAYES HOTEL

It was in the billiards room of the still-functioning Hayes Hotel at 3pm on 1 November 1884 that seven men led by Michel Cusack opened a meeting and established the Gaelic Athletic Association, usually referred to since as the GAA or as 'the Gaaaaa'. 130-plus years on and the GAA now numbers more than a million members in Ireland and abroad, from Westmeath to the west coast of the US, Abbeyknockmoy to Abu Dhabi, Ballyhaunis to Bangkok

.

SEMPLE STADiUM

While Croke Park is the home of football and annually hosts the All-Ireland finals of both codes, it is the sweltering cauldron of Semple Stadium that often plays host to the most hotly contested hurling matches of the year. With an attendance of some 53,000, Semple is the GAA's second largest stadium and the traditional home of Munster hurling, where the province's best go at it every July to see who will go forth as Munster champions. And it is here on those hot summer days, when even the grass seems like it is about to catch fire, that some of the country's most epic hurling contests have taken place.

These classics include the 1991 replay between Tipp and Cork, when Pat Fox's goal sparked a pitch invasion led by a Tipp fan in a wheelchair; the 2004 final, when despite being reduced to fourteen men, against the wind and trailing, Waterford rallied to beat the Rebels; and perhaps Semple Stadium's most famous epic, the 1994 clash between Californian rivals Rage Against the Machine and Cypress Hill. However, because this final contest wasn't officially part of the Munster Hurling Championship but was instead during the now legendary Féile musical festivals that took place in Semple Stadium during the early 1990s, it was Limerick and not Cypress Hill that went on to represent Munster in the All-Ireland semis that year.

WATERFORD

Located on Ireland's sunny south-east Waterford, or the Déise as it is commonly called, has a greater abundance of tourist sites than it is sometimes given credit for. From the cosmopolitan centre of Waterford city in the east to the welcoming atmosphere of the Ring Gaeltacht in the west, and from walking along the Comeragh Mountains of the north to lazing about any number of the secluded beaches on the Copper Coast in the south, Waterford is well-stocked with reasons to visit.

That said, Waterford's past does have a darker side, with several things that it will probably have to be held account-able for on Judgement Day, including:

1. NEWFOUNDLAND'S ACCENT: Although residents of Newfoundland, Canada, usually get a hard time from main-land Canadians due to their isolation on an island off Canada's east coast, their accent doesn't help matters. A traditional 'Newfie' accent sounds like a Waterford native on steroids, and this is due to the large numbers of Déise fish-ermen and families who emigrated over in the 1800s, boy!

2. €70 FEES FOR PRINTING BOARDING PASSES:
Ryanair's first flight was a swift hop from Waterford to Gatwick airport. The rest is history.

3. THE GLOBAL DOMINANCE OF THE NEW ZEALAND RUGBY TEAM:
Ireland has never beaten the All-Blacks and this is because of Waterford. Back in 1840, Déise-born Captain William Hobson co-authored the Treaty of Waitangi between Great Britain and the Maoris of New Zealand. This document helped establish trust between both groups, arguably making the Maoris more amenable to taking part in English games such as rugby, which is a pity because they are so bloody good at it!

4. CORPORAL PUNISHMENT:
The first Christian Brothers school was established in Waterford, which, as any young teenage boy who has had his locks pulled or a chalk-duster thrown at him to grab his attention will tell you, is not entirely something to cheer about.

5. CARDIAC ARRESTS:
While it may be a little strong to blame heart attacks on one county, the Déise did come up with the curing process for bacon, when Henry Denny developed the modern-day rasher back in the early 19th century. Irish breakfasts and cholesterol levels have never been the same since.

6. CELEBRITY WEDDINGS: The forerunner for celebrity weddings took place in Waterford on 29 August 1170 when Strongbow married Aoife. This union between the infamous Norman leader and the daughter of an Irish lord not only gave Strongbow succession rights to the kingdom of Leinster, it lay the template for the lavish ceremonies we have today. It is still a secret how much *Hello* paid the couple for exclusive photos of the event.

7. NUCLEAR APOCALYPSE?: Ernest Walton. Dungarvan native. Nobel Prize winner in physics. Helped usher in the nuclear age by being the first person to split the atom – need I say more?

TRAMORE METAL MAN

The Tramore Metal Man, a four-metre-high metal statue of an Ancient Mariner pointing seawards, stands proudly on top of one of three pillars on the cliffs near the coastal town of Tramore. Legend has it that single ladies need only to hop bare-footed around the Metal Man's pillar three times and they will be married within the year. This puts the Tramore Metal Man just behind Tinder and immediately in front of finding a ring in your barmbrack on the list of ways women can find their future husbands.

Although single women everywhere greatly appreciate that the Metal Man has been doing his bit as Cupid since his construction in 1823, the real reason for his position on the Tramore coastline is as a warning to passing seafarers of the treacherous sea conditions nearby.

By topographical coincidence, the bay of Tramore, which is renowned for its peril, looks remarkably like the mouth of Waterford harbour and on more than one occasion ships caught in storms and squalls, unable to spot the lights and

guidance offered by Hook Lighthouse, were drawn into Tramore Bay and were unable to escape.

After the wrecking of transport vessel the Sea Horse in 1816, with the loss of 360 lives, measures were taken to help prevent any further tragedy, leading to the construction of the Metal Man. While another identical Metal Man exists at the entrance to Sligo harbour, Tramore's Metal Man's position, on top of one of three huge eighteen-metre-high stone pillars, makes him that little bit more impressive.

So for nearly 200 years since, despite looking like a slightly camp 18th-century member of the Village People, Tramore's Ancient Mariner has proudly pointed seawards warning seafarers away from the bay's dangerously shallow waters and the cliff rocks directly below. And it is said that even today, despite the advent of GPS and the decline of wooden sailing ships, the Metal Man still toils away and can be heard chanting during the worst south-easterly storms, '*keep out, keep out, good ship, from me, for I am the rock of misery*'.

DUNMORE EAST

While this book can't speak for Dunmore North, South or West, it can say that of all our charming Irish fishing villages, Dunmore East is one of the finest. Divided into an upper and lower village, with a sheltered beach, colourful harbour, scattering of pretty thatched cottages and fine shops, pubs and restaurants, its reputation as one of the south-east's most inviting seaside villages is richly deserved.

While the village makes a popular and peaceful retreat all year round, it is good to be aware that during the warmer months, the commotion and racket that often emanates down by the harbour is not a Club 18–30 sun holiday that has rocked up to party but the resident colony of noisy kittiwake seabirds that has forsaken their kind's usual breeding habit of avoiding people and have chosen to roost right beside the village.

ARDMORE OGHAM STONES

Ogham, spelt with a silent 'g' and pronounced 'ohm' (just imagine you've been kicked lightly in the balls), is an exceptionally old medieval alphabet that was used to write the early Irish language until the 6th century, when the Roman alphabet arrived in. Ogham writing is a series of scratches that, to the untrained eye, looks like somebody counting down the days to Christmas – and doing it badly.

To the trained eye, however, ogham is a complex language that uses four groups of five letters, thus creating an alphabet of 20 symbols, each of which also refers to a shrub or tree.

Today, we know roughly how ogham works. However, what we still don't know is why ogham was used in the first place, but here are a few possible suggestions.

- One theory is that ogham was a Celtic version of Germanic runes (their version of scratchy symbols).

- Another idea is that ogham was our way of excluding the Romans from our earliest communications – even though there were no Romans around to listen in to our earliest communications!

- A third theory argues that ogham was some sort of religious communication between the druids, most of which took the form of scratching pieces of wood with ogham symbols and then sending them to each other, in effect creating a kind of tree-mail.

- Another contends that ogham came from the Greek alphabet and started out as a series of hand signals before evolving into written form.

- A fifth, inspired by the 11th-century Irish history book *Lebor Gabála Érenn*, says ogham came from the Tower of Babel (where all languages supposedly originate) when Scythian king Fenius Farsa took the best bits from every language to make the best language ever and ogham was created to be the perfect writing system for it.

- And finally, a slightly more far-fetched idea is that ogham was created to help guide the way towards secret treasure, the evidence for which is the 2003 *Murder She Wrote* TV movie *The Celtic Riddle*, which is set in Kerry and where ogham maps are used to deduce not only whodunit but the location of secret treasure. Television gold.

Today, most of our remaining examples of ogham come from standing stones that either act as memorials to someone who died or to demarcate land boundaries, with titles usually inscribed on them that can be read from bottom to top. While these ogham stones can be found throughout the country, for some reason they are most frequent in the south-west.

Of the many that can be found, two hugely impressive examples are situated in the ecclesiastic ruins of Ardmore, which coincidentally just happens to be the first Christian settlement in Ireland.

THE COPPER COAST

Long before man learnt to burn electrical wire in a remote bog to make copper, he had much more interesting and fun ways to extract it. Remnants of these ways can still be seen along Waterford's stunning Copper Coast, so named due to the mining activity that took place here in the mid-1800s and the copperish hue that characterises its cliff faces. For a time, one of the most important mining districts in the British Empire, the Copper Coast has quickly become one of the south-east's primary scenic attractions.

Having been declared a UNESCO Global Geopark in 2004, the area, which stretches along the coast from Stradbally to Fenor, is ideal for walking, cycling and discovering small secluded coves and beaches. While it doesn't really matter in which direction you travel the Copper Coast, don't panic if, when you reach Fenor, you see throngs of men, women and children running down the road towards you. They're not fleeing a south-eastern tsunami but partaking in a round of the local road-bowling that you occasionally find down here and in other pockets of the island.